D1090764

Differential Equations with Small Parameters and Relaxation Oscillations

MATHEMATICAL CONCEPTS AND METHODS IN SCIENCE AND ENGINEERING

Series Editor: **Angelo Miele**
Mechanical Engineering and Mathematical Sciences, Rice University

1 **INTRODUCTION TO VECTORS AND TENSORS** • Volume 1: Linear and Multilinear Algebra • *Ray M. Bowen and C.-C. Wang*

2 **INTRODUCTION TO VECTORS AND TENSORS** • Volume 2: Vector and Tensor Analysis • *Ray M. Bowen and C.-C. Wang*

3 **MULTICRITERIA DECISION MAKING AND DIFFERENTIAL GAMES** • *Edited by George Leitmann*

4 **ANALYTICAL DYNAMICS OF DISCRETE SYSTEMS** • *Reinhardt M. Rosenberg*

5 **TOPOLOGY AND MAPS** • *Taqdir Husain*

6 **REAL AND FUNCTIONAL ANALYSIS** • *A. Mukherjea and K. Pothoven*

7 **PRINCIPLES OF OPTIMAL CONTROL THEORY** • *R. V. Gamkrelidze*

8 **INTRODUCTION TO THE LAPLACE TRANSFORM** • *Peter K. F. Kuhfittig*

9 **MATHEMATICAL LOGIC:** An Introduction to Model Theory • *A. H. Lightstone*

11 **INTEGRAL TRANSFORMS IN SCIENCE AND ENGINEERING** • *Kurt Bernardo Wolf*

12 **APPLIED MATHEMATICS:** An Intellectual Orientation • *Francis J. Murray*

13 **DIFFERENTIAL EQUATIONS WITH SMALL PARAMETERS AND RELAXATION OSCILLATIONS** • *E. F. Mishchenko and N. Kh. Rozov*

14 **PRINCIPLES AND PROCEDURES OF NUMERICAL ANALYSIS** • *Ferenc Szidarovszky and Sidney Yakowitz*

16 **MATHEMATICAL PRINCIPLES IN MECHANICS AND ELECTROMAGNETISM,** Part A: Analytical and Continuum Mechanics • *C.-C. Wang*

17 **MATHEMATICAL PRINCIPLES IN MECHANICS AND ELECTROMAGNETISM,** Part B: Electromagnetism and Gravitation • *C.-C. Wang*

18 **SOLUTION METHODS FOR INTEGRAL EQUATIONS:** Theory and Applications • *Edited by Michael A. Golberg*

19 **DYNAMIC OPTIMIZATION AND MATHEMATICAL ECONOMICS** • *Edited by Pan-Tai Liu*

20 **DYNAMICAL SYSTEMS AND EVOLUTION EQUATIONS:** Theory and Applications • *J. A. Walker*

21 **ADVANCES IN GEOMETRIC PROGRAMMING** • *Edited by Mordecai Avriel*

Differential Equations with Small Parameters and Relaxation Oscillations

E.F. Mishchenko and N.Kh. Rozov

Academy of Sciences of the USSR
Moscow, USSR

Translated from Russian by
F. M. C. Goodspeed

PLENUM PRESS · NEW YORK AND LONDON

Library of Congress Cataloging in Publication Data

Mishchenko, Evgenii Frolovich.
Differential equations with small parameters and relaxation oscillations.

Translation of Differentsial'nye uravneniia s malym parametrom i relaksatsionnye kolebaniia.
Bibliography: p.
Includes index.
1. Differential equations–Numerical solutions. 2. Differential equations–Asymptotic theory. 3. Relaxation methods (Mathematics). I. Rozov, Nikolai Khristovich, joint author. II. Title.
QA371.M5913 515'.35 78-4517
ISBN 0-306-39253-4

The original Russian text, published by Nauka Press in Moscow in 1975, has been corrected by the authors for the present edition. This translation is published under an agreement with the Copyright Agency of the USSR (VAAP).

Дифференциальные уравнения с малым параметром
и релаксационные колебания

DIFFERENTSIAL'NYE URAVNENIYA S MALYM PARAMETROM
I RELAKSATSIONNYE KOLEBANIYA
E. F. Mishchenko and N. Kh. Rozov

A Division of Plenum Publishing Corporation
227 West 17th Street, New York, N.Y. 10011

Printed in the United States of America

PREFACE

A large amount of work has been done on ordinary differential equations with small parameters multiplying derivatives. This book investigates questions related to the asymptotic calculation of relaxation oscillations, which are periodic solutions formed of sections of both slow- and fast-motion parts of phase trajectories. A detailed discussion of solutions of differential equations involving small parameters is given for regions near singular points.

The main results examined were obtained by L. S. Pontryagin and the authors. Other works have also been taken into account: A. A. Dorodnitsyn's investigations of Van der Pol's equation, results obtained by N. A. Zheleztsov and L. V. Rodygin concerning relaxation oscillations in electronic devices, and results due to A. N. Tikhonov and A. B. Vasil'eva concerning differential equations with small parameters multiplying certain derivatives.

E. F. Mishchenko
N. Kh. Rozov

CONTENTS

Chapter I. Dependence of Solutions on Small Parameters. Applications of Relaxation Oscillations

1. Smooth Dependence. Poincaré's Theorem 1

2. Dependence of Solutions on a Parameter, on an Infinite Time Interval 3

3. Equations with Small Parameters Multiplying Derivatives 4

4. Second-Order Systems. Fast and Slow Motion. Relaxation Oscillations 9

5. Systems of Arbitrary Order. Fast and Slow Motion. Relaxation Oscillations 16

6. Solutions of the Degenerate Equation System 24

7. Asymptotic Expansions of Solutions with Respect to a Parameter 29

8. A Sketch of the Principal Results 34

Chapter II. Second-Order Systems. Asymptotic Calculation of Solutions

1. Assumptions and Definitions 39

2. The Zeroth Approximation 45

3. Asymptotic Approximations on Slow-Motion Parts of the Trajectory 48

4. Proof of the Asymptotic Representations of the Slow-Motion Part 52

5. Local Coordinates in the Neighborhood of a Junction Point 56

6. Asymptotic Approximations of the Trajectory on the Initial Part of a Junction 60

7. The Relation between Asymptotic Representations and Actual Trajectories in the Initial Junction Section 62

8. Special Variables for the Junction Section 67

9. A Riccati Equation 68

10. Asymptotic Approximations for the Trajectory in the Neighborhood of a Junction Point 72

11. The Relation between Asymptotic Approximations and Actual Trajectories in the Immediate Vicinity of a Junction Point 77

12. Asymptotic Series for the Coefficients of the Expansion Near a Junction Point 82

13. Regularization of Improper Integrals 89

14. Asymptotic Expansions for the End of a Junction Part of a Trajectory 96

15. The Relation between Asymptotic Approximations and Actual Trajectories at the End of a Junction Part 99

16. Proof of Asymptotic Representations for the Junction Part 103

17. Asymptotic Approximations of the Trajectory on the Fast-Motion Part 107

18. Derivation of Asymptotic Representations for the Fast-Motion Part 111

19. Special Variables for the Drop Part 114

20. Asymptotic Approximations of the Drop Part of the Trajectory 119

21. Proof of Asymptotic Representations for the Drop Part of the Trajectory 125

22. Asymptotic Approximations of the Trajectory for Initial Slow-Motion and Drop Parts 132

Chapter III. Second-Order Systems. Almost-Discontinuous Periodic solutions

1. Existence and Uniqueness of an Almost-Discontinuous Periodic Solution 139

2. Asymptotic Approximations for the Trajectory of a Periodic Solution 143

3. Calculation of the Slow-Motion Time 144

4. Calculation of the Junction Time 145

5. Calculation of the Fast-Motion Time 156

6. Calculation of the Drop Time 157

7. An Asymptotic Formula for the Relaxation-Oscillation Period 164

8. Van der Pol's Equation. Dorodnitsyn's Formula 168

Chapter IV. Systems of Arbitrary Order. Asymptotic Calculation of Solutions

1. Basic Assumptions 171

2. The Zeroth Approximation 173

3. Local Coordinates in the Neighborhood of a Junction Point 176

4. Asymptotic Approximations of a Trajectory at the Beginning of a Junction Section 179

5. Asymptotic Approximations for the Trajectory in the Neighborhood of a Junction Point 187

6. Asymptotic Approximation of a Trajectory at the End of a Junction Section 193

7. The Displacement Vector 196

Chapter V. Systems of Arbitrary Order. Almost-Discontinuous Periodic Solutions

1. Auxiliary Results 199

2. The Existence of an Almost-Discontinuous Periodic Solution. Asymptotic Calculation of the Trajectory 203

3. An Asymptotic Formula for the Period of Relaxation Oscillations 209

References 217

Index 223

CHAPTER I

DEPENDENCE OF SOLUTIONS ON SMALL PARAMETERS. APPLICATIONS OF RELAXATION OSCILLATIONS

When the operation of a device or the course of a process is described by differential equations, we are passing from an actual object (process) to an idealized model. Every mathematical idealization involves, to a certain extent, the neglect of small quantities. Hence the question of how much distortion of the original phenomenon is introduced becomes important. We thus arrive at the mathematical problem of the dependence of solutions of differential equations on small parameters.

In this chapter we consider general characteristics of various types of dependence in the case of a normal autonomous system of ordinary differential equations. To simplify the exposition we consider problems involving only one parameter.

1. Smooth Dependence. Poincaré's Theorem

We consider the autonomous differential equation system

$$\dot{x}^i = F^i(x^1, \ldots, x^n, \varepsilon), \quad i = 1, \ldots, n, \tag{1.1}$$

or, in vector form, the equation

$$\dot{x} = F(x, \varepsilon), \tag{1.2}$$

where $x = (x^1, \ldots, x^n)$ is an n-vector of euclidean vector space R^n, $F(x, \varepsilon) = (F^1(x, \varepsilon), \ldots, F^n(x, \varepsilon))$ is an n-dimensional vector-valued function of x and ε, and ε is a numerical parameter. We first assume that ε is *small*,

$$0 \leqslant \varepsilon \leqslant \varepsilon_0, \tag{1.3}$$

where ε_0 is a small number.

Let the functions $F^i(x^1, \ldots, x^n)$, $i = 1, \ldots, n$ be defined and *continuous* in some domain G of the variables $x^1, \ldots, x^n$, where ε satisfies (1.3). We write

$$x = \varphi(t, \varepsilon) \tag{1.4}$$

for the solution of (1.2) satisfying the initial condition $x_0 = \varphi(t_0, \varepsilon)$, $(x_0, \varepsilon) \in G$. Together with (1.2) we consider the system

$$\dot{x} = F(x, 0), \tag{1.5}$$

obtained from (1.2) by putting $\varepsilon = 0$. Let

$$x = \varphi_0(t) \tag{1.6}$$

be the solution of (1.5) with the same initial condition $x_0 = \varphi_0(t_0)$, defined on some *finite* time interval

$$t_0 \leqslant t \leqslant T. \tag{1.7}$$

If ε is small, the right sides in (1.2) and (1.5) differ by only a small quantity. It is natural to ask how the solutions (1.4) and (1.6) differ. In many cases important in practice, this question is answered by the following well-known theorems [45, 30, 23, 3].

Theorem 1 (concerning the continuous dependence of solutions on a parameter). If the right sides in (1.2) are continuously differentiable with respect to $x^1, \ldots, x^n$ and continuous functions of ε in a region G, then, for sufficiently small ε, the solution (1.4) is defined on the same interval (1.7) as the solution (1.6), and

$$\varphi(t, \varepsilon) = \varphi_0(t) + R_0(t, \varepsilon), \tag{1.8}$$

where $R_0(t, \varepsilon) \to 0$ when $\varepsilon \to 0$, uniformly with respect to t on the interval (1.7).

Theorem 2 (concerning the differentiability of solutions with respect to a parameter). If the right sides in (1.2) have, in G, continuous partial derivatives up to order $m \geq 1$, inclusive, with respect to the totality of all arguments, then, if ε is small enough, the solution (1.4) has the representation

$$\varphi(t, \varepsilon) = \varphi_0(t) + \varepsilon\varphi_1(t) + \ldots + \varepsilon^{m-1}\varphi_{m-1}(t) + R_m(t, \varepsilon), \tag{1.9}$$

where $R_m(t, \varepsilon) \to 0$, when $\varepsilon \to 0$ like ε^m, uniformly with respect to t on the interval (1.7).

Theorem 3 (Poincaré's theorem on the analyticity of solutions as functions of a parameter). If the right sides in (1.2) are analytic functions of each of their arguments in G, then, for sufficiently small ε, a solution (1.4) has the representation

$$\varphi(t, \varepsilon) = \varphi_0(t) + \sum_{m=1}^{\infty} \varepsilon^m \varphi_m(t); \tag{1.10}$$

the series converges uniformly on the interval (1.7).

Theorems 1, 2, and 3 not only confirm that, for small but *finite* ε, the solution (1.4) differs only slightly from the solution (1.6), but also indicate a method of finding this difference with any required degree of accuracy.

2. Dependence of Solutions on a Parameter, on an Infinite Time Interval

Theorems 1, 2, and 3 give no answer to the question of the deviation of the solution (1.4) from the solution (1.6) on an *infinite* time interval. Simple examples show that this deviation is not always small. Moreover, even if the solution (1.6) is defined for all $t \geq t_0$, the solution (1.4) is not always defined for all $t \geq t_0$.

Example 1. The scalar equation

$$\dot{x} = (x + \varepsilon)^2 \tag{2.1}$$

becomes

$$\dot{x} = x^2 \tag{2.2}$$

when $\varepsilon = 0$. The solution of (2.2) with zero initial value for $t = 0$ is $x = \varphi_0(t) \equiv 0$, $0 \leq t < \infty$, while the solution of (2.1) with the same initial values is

$$x = \varphi(t, \varepsilon) \equiv \frac{\varepsilon}{1 - \varepsilon t} - \varepsilon;$$

this solution is defined only for $0 \leq t < 1/\varepsilon$.

Example 2. Consider an electric circuit, formed of a condenser with capacitance C and a coil with inductance L in series (Fig. 1). If we neglect the small resistance of the circuit, the dependence of the current i on the time is described by the following equation [45]:

$$L \frac{d^2 i}{dt^2} + \frac{1}{C} i = 0. \tag{2.3}$$

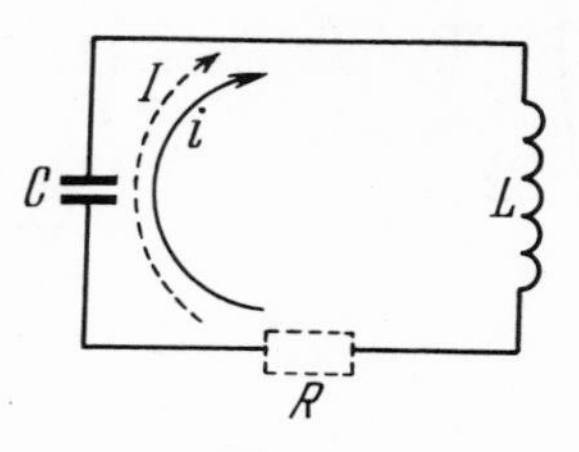

Fig. 1

But is this idealization, with the resistance R neglected, justified? In other words, does the solution of (2.3) differ only slightly from the solution of the equation

$$L\frac{d^2I}{dt^2}+R\frac{dI}{dt}+\frac{1}{C}I=0 \tag{2.4}$$

when R is small?

If we are interested only in a finite time interval, solutions of (2.3) and (2.4) with the same initial conditions differ only by a small quantity. On an infinite time interval, however, this is not true; in fact $I(t) \to 0$ when $t \to \infty$, but $i(t)$ performs periodic oscillations with constant amplitude. The phase portraits of Eqs. (2.3) and (2.4) in the $(i, di/dt)$ and the $(I, dI/dt)$ plane differ strikingly; the only equilibrium position for (2.3) is a center, while Eq. (2.4) has a focus (Fig. 2).

3. Equations with Small Parameters Multiplying Derivatives

Another reason for Theorems 1, 2, and 3 to be inapplicable for estimating the deviation of the solution (1.4) from the solution (1.6), even on a finite time interval, is a *discontinuity* (or lack of smoothness) in the dependence of the right sides in (1.1) on ε. This occurs in normal systems in which a small positive parameter ε occurs as a coefficient of some derivatives, for example in the system

$$\begin{cases} \varepsilon\dot{x}^i=f^i(x^1, \ldots, x^k, y^1, \ldots, y^l), & i=1, \ldots, k, \\ \dot{y}^j=g^j(x^1, \ldots, x^k, y^1, \ldots, y^l), & j=1, \ldots, l, \end{cases} \tag{3.1}$$

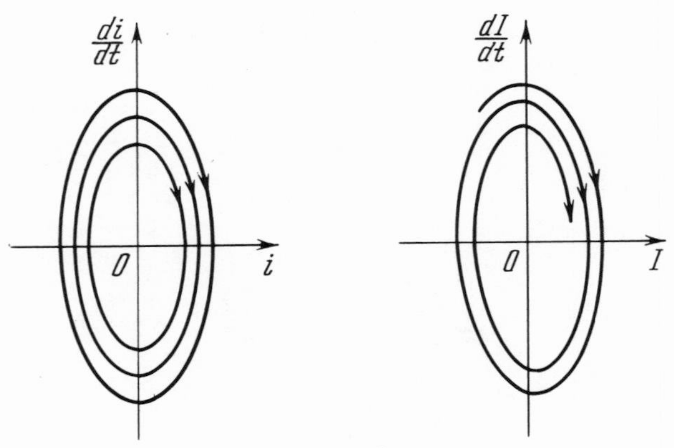

Fig. 2

where f^i and g^j are smooth functions of all their $k + l = n$ arguments. It is clear that, if (3.1) is written in the form (1.1), the right sides contain functions $(1/\varepsilon)f^i$, which increase without limit when $\varepsilon \to 0$.

We can reduce (3.1) to the form (1.1), so that the right side is a smooth function of ε. To this end, we put $t = \varepsilon\theta$ to obtain

$$\begin{cases} \dfrac{dx^i}{d\theta} = f^i(x^1, \ldots, x^k, y^1, \ldots, y^l), & i = 1, \ldots, k, \\ \dfrac{dy^j}{d\theta} = \varepsilon g^j(x^1, \ldots, x^k, y^1, \ldots, y^l), & j = 1, \ldots, l. \end{cases} \tag{3.2}$$

Theorems 1, 2, and 3 can be applied to (3.2), but this is not of great practical interest. We can ensure that solutions of (3.2) and solutions of the system obtained from it by putting $\varepsilon = 0$ differ by a small amount only on a finite interval of values of θ, i.e., on a time interval *whose length tends to zero* with ε.

We next consider two physical situations described by differential equation systems of the form (3.1). These examples will also be useful later in the illustration of certain phenomena.

Example 3 (Van der Pol's equation). Consider a vacuum tube oscillator consisting of a triode with an oscillating anode circuit; the circuit diagram is shown in Fig. 3. If I is the current passing through the resistance ba or, equivalently, through the inductance kb, then I, as a function of t, is described by the following differential equation

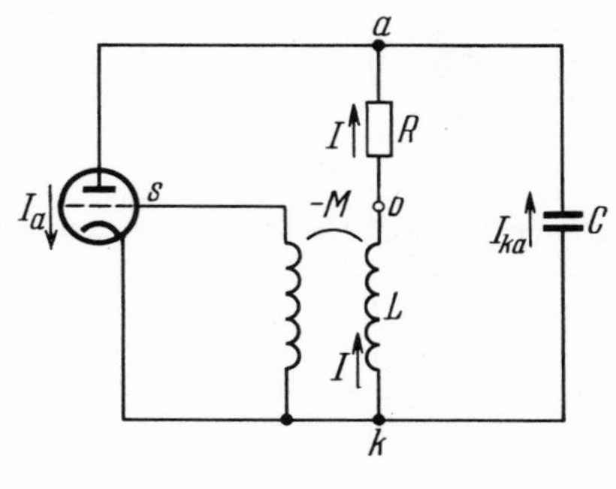

Fig. 3

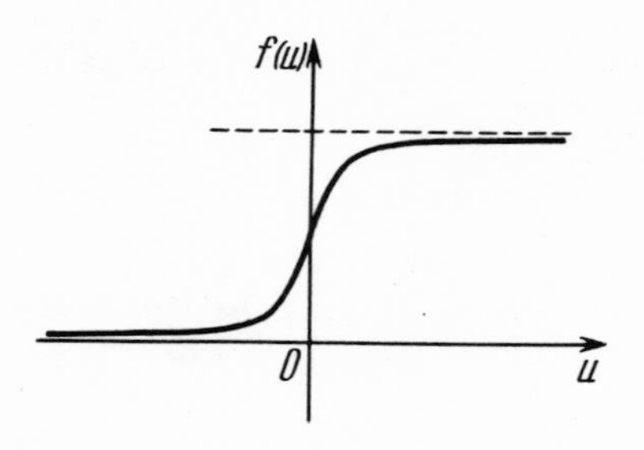

Fig. 4

[45, 7, 3]:

$$L\frac{d^2I}{dt^2}+R\frac{dI}{dt}+\frac{1}{C}I=\frac{1}{C}f\left(M\frac{dI}{dt}\right); \tag{3.3}$$

here M is a positive constant (the mutual inductance) and $f(u)$, a characteristic of the vacuum tube, is a smooth monotonically increasing function of u with a graph as shown in Fig. 4. We can assume that $f'(u)$ has its maximum value for $u = 0$, i.e., $f''(0) = 0$ and $f'''(0) < 0$.

It is known [45] that a vacuum tube oscillator generates undamped periodic oscillations if its parameters satisfy the condition

$$R<\frac{M}{C}f'(0); \tag{3.4}$$

in this case, Eq. (3.3) has a single stable limit cycle in the $(I, dI/dt)$ plane. If $I(t)$ is replaced by the unknown function $i(t) = I(t) - f(0)$, Eq. (3.3) becomes *Rayleigh's equation*

$$LC\frac{d^2i}{dt^2}+F\left(\frac{di}{dt}\right)+i=0, \tag{3.5}$$

where $F(v) \equiv RCv - f(Mv) + F(0)$.

We now consider the idealized case in which $f(u)$ can (at least when the absolute value of u is not too large) be replaced by a cubic polynomial

$$f(u)=f(0)+f'(0)u+\frac{1}{6}f'''(0)u^3;$$

then

$$F(v)=(RC-f'(0)M)v-\frac{1}{6}f'''(0)M^3v^3,$$

$f'''(0)M^3 < 0$, $RC - f'(0)M < 0$ by virtue of condition (3.4), and (3.5) becomes

$$LC\frac{d^2i}{dt^2} + \left[(RC - f'(0)M) - \frac{1}{6}f'''(0)M^3\left(\frac{di}{dt}\right)^2\right]\frac{di}{dt} + i = 0.$$

If we transform to the new time $t = t/(LC)^{\frac{1}{2}}$ and the new unknown

$$z = \alpha i, \text{ where } \alpha^2 = \frac{RC - f'(0)M}{\frac{1}{2}f'''(0)M^3}LC,$$

then we obtain the equation

$$\frac{d^2z}{dt^2} + \lambda\left[-\frac{dz}{dt} + \frac{1}{3}\left(\frac{dz}{dt}\right)^3\right] + z = 0,$$

where $\lambda = \dfrac{f'(0)M - RC}{\sqrt{LC}} > 0.$

Finally, differentiating this last equation with respect to t and putting $x = dz/dt$, we obtain *Van der Pol's equation* [11]

$$\frac{d^2x}{dt^2} + \lambda[-1 + x^2]\frac{dx}{dt} + x = 0. \tag{3.6}$$

Equation (3.6) describes the operation of a vacuum tube oscillator in our idealization. The parameters of the generator are characterized by the single parameter λ. We have already noted that, if (3.4) holds, we have self-excited periodic oscillations (auto-oscillations); mathematically, this corresponds to the fact that Van der Pol's equation with any $\lambda > 0$ has a stable limit cycle in the $(x, dx/dt)$ plane.

For small λ, Eq. (3.6) differs only slightly from the equation for a linear oscillator, and the auto-oscillations of the generator are close to simple harmonic oscillations. As λ increases, the auto-oscillations differ more and more from harmonic oscillations, and for large λ they take the essentially different form of *relaxation oscillations* (see Sec. 4).

For large $\lambda > 0$, (3.6) can easily be reduced to the form (3.1). In fact, if we put

$$y = \int_0^x (x^2 - 1)\,dx + \frac{1}{\lambda}\frac{dx}{dt}, \quad t_1 = \frac{t}{\lambda}, \quad \varepsilon = \frac{1}{\lambda^2}, \tag{3.7}$$

then after the obvious transformations we obtain from (3.6) a second-order system (in which we write t instead of t_1 for simplicity)

$$\begin{cases} \varepsilon \dfrac{dx}{dt} = y - \dfrac{1}{3}x^3 + x, \\ \dfrac{dy}{dt} = -x; \end{cases} \tag{3.8}$$

here $\varepsilon > 0$ is a *small* parameter. We shall also refer to (3.8) as a Van der Pol equation.

Second-order systems of the form (3.1) (i.e., with $k = l = 1$) arise in the study of many electronic devices (for example, vacuum tube multivibrators with a single RC-component), in the description of the operation of which small parasitic capacitances play an important role [3], and in some types of multivibrators using tunnel diodes [54].

Example 4. The operation of several electronic devices (for example, two-tube Frugauer generators and symmetric multivibrators [3]), when small parasitic capacitances and inductances are taken into account, is described by a fourth-order differential system of the form

$$\begin{cases} \varepsilon \dot{x}^1 = -\alpha(y^1 - y^2) + \varphi(x^1) - x^2, \\ \varepsilon \dot{x}^2 = \alpha(y^1 - y^2) + \varphi(x^2) - x^1, \\ \dot{y}^1 = x^1, \\ \dot{y}^2 = x^2; \end{cases} \tag{3.9}$$

here $\alpha > 0$ is a constant and $\varphi(u)$, $-1 < u < 1$, is a function of u whose graph is shown in Fig. 5 (which is obtained by a transformation of a vacuum tube characteristic function; see Fig. 4). If the parasitic parameters are not taken into account, the operation of these devices is described by the equations obtained by putting $\varepsilon = 0$ in (3.9):

$$\begin{cases} -\alpha(y^1 - y^2) + \varphi(x^1) - x^2 = 0, \\ \alpha(y^1 - y^2) + \varphi(x^2) - x^1 = 0, \\ \dot{y}^1 = x^1, \\ \dot{y}^2 = x^2. \end{cases} \tag{3.10}$$

Radiophysicists have recently discovered that such devices can generate periodic oscillations of an unusual nature: At certain times (or for certain current strengths), discontinuous changes can occur between periods of smooth variations.

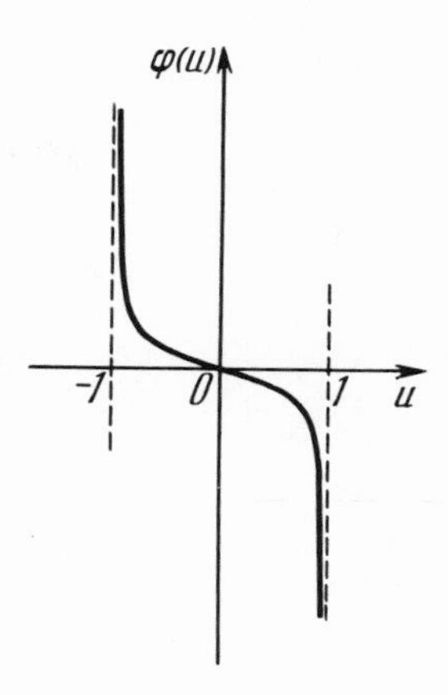

Fig. 5

Oscillations of this type are called *relaxation* oscillations. However all attempts at a theoretical explanation of this phenomenon by means of the system (3.10) failed. It was found necessary to introduce supplementary physical hypotheses (the "discontinuity hypothesis" [27]). The first purely mathematical explanation of relaxation oscillation in systems of the form (3.1), without any extra physical hypotheses, was given in [20] and developed further in [38]. We discuss this explanation in Sec. 5, using the systems (3.9) and (3.10) as examples.

4. Second-Order Systems. Fast and Slow Motion. Relaxation Oscillations

We now consider the second-order system

$$\begin{cases} \varepsilon\dot{x} = f(x, y), \\ \dot{y} = g(x, y), \end{cases} \tag{4.1}$$

where x and y are scalar functions of t and ε is a small positive parameter. Let

$$\begin{cases} f(x, y) = 0, \\ \dot{y} = g(x, y) \end{cases} \tag{4.2}$$

be the *degenerate system* corresponding to (4.1), i.e., the system obtained from (4.1) by putting $\varepsilon = 0$. The system (4.2) is not a normal differential equation system [the first of (4.2) is not a differential equation]. Hence it does not have solutions with arbitrary initial values (x_0, y_0). We must find solutions with initial points on the curve $f(x,y) = 0$, since all trajectories of the degenerate system (4.2) are on this curve (by virtue of the first equation).

Thus the question of the difference between solutions of (4.1) and (4.2) has a meaning only for solutions of (4.1) beginning at points of a small neighborhood of the curve $f(x,y) = 0$ (small together with ε). However, even these solutions of (4.1) do not always tend to solutions of (4.2) when $\varepsilon \to 0$. It is also important to find conditions under which any trajectory of (4.1), starting at an initial value (x_0,y_0) at a finite distance from the curve $f(x,y) = 0$, reaches a small (together with ε) neighborhood of this curve, the time taken in reaching this neighborhood, etc.

We consider these questions in detail in Chapters II and III. Here we describe only some of the results (without proofs or precise estimates); we try to give a clear explanation of their origin and, in particular, we show that (4.1) can have periodic solutions of the nature of relaxation oscillations.

We start with Van der Pol's equation (3.8), for which the corresponding degenerate system is

$$\begin{cases} y-\frac{1}{3}x^3+x=0, \\ \dot{y}=-x; \end{cases} \tag{4.3}$$

hence trajectories of all solutions of the degenerate system are on the cubic parabola

$$y-\frac{1}{3}x^3+x=0. \tag{4.4}$$

The system (4.3) has five complete trajectories (Fig. 6):

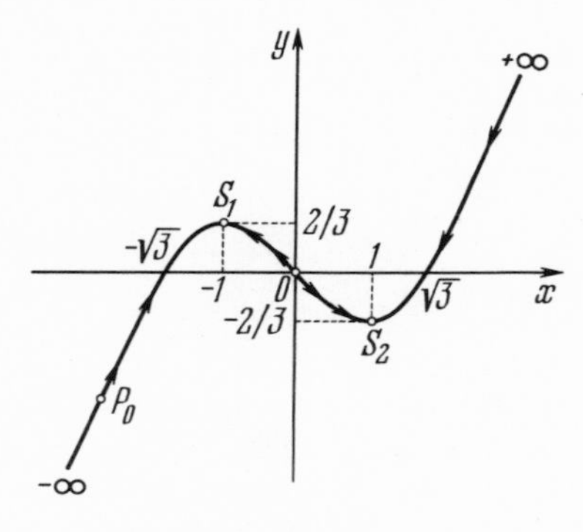

Fig. 6

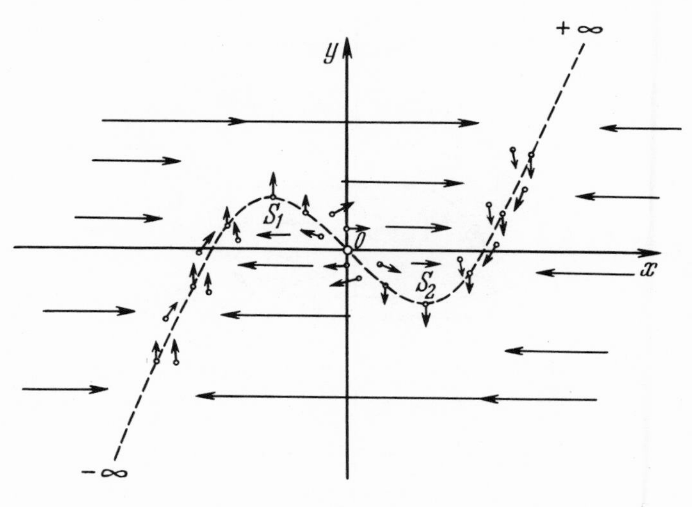

Fig. 7

$(-\infty, S_1)$, $(+\infty, S_2)$, $(0, S_1)$, $(0, S_2)$ (the directions of each of these trajectories for increasing t are shown by arrows), and the point O, which is the unique position of stability. An important fact is that the phase point of the system (4.3) starting from a point P_0, for example, on the branch $(-\infty, S_1)$ of the curve (4.4), reaches the point S_1 in a *finite* time. However there is no trajectory of (4.3) leaving S_1; hence no conclusion can be reached by considering the degenerate equation concerning further motion of the phase point.

On the other hand, the vector field of the phase velocity of the nondegenerate system (3.8) is easily constructed (Fig. 7); it shows that a trajectory of (3.8), starting from an arbitrary initial point Q_0 not on the curve (4.4), first enters a small (together with ε) neighborhood of the branch $(-\infty, S_1)$ [or the branch $(+\infty, S_2)$], and then, for all values of the time, passes close to the contour $Z_0 = P_2S_1P_1S_2$ formed of the horizontal segments S_1P_1, S_2P_2 and the arcs P_2S_1, P_1S_2 of the curve (4.4) (Fig. 8).

It is thus clear that, near Z_0, there is a closed trajectory Z_ε of the system (3.8) (Fig. 9). In fact, in Fig. 8, consider a short segment L_1L_2 parallel to the Ox axis, cutting the arc P_2S_1 at an interior point L. The foregoing implies that this segment is mapped, by trajectories of (3.8), into a part of itself at a distance from L that is small when ε is small. This mapping has a fixed point through which a closed trajectory of (3.8) passes.

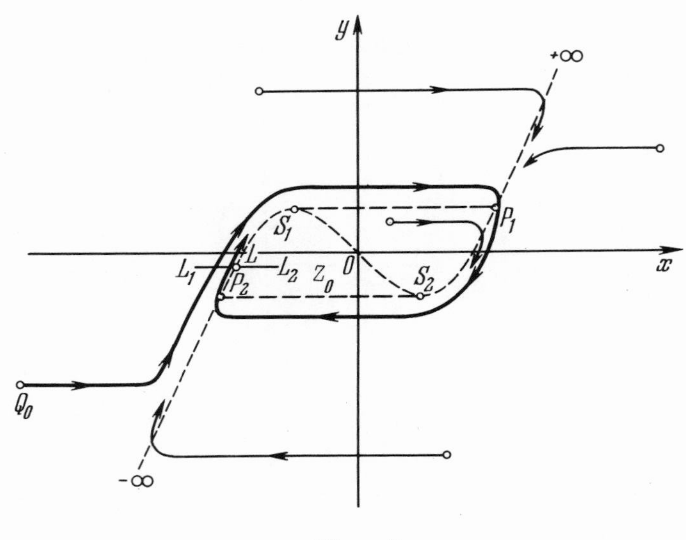

Fig. 8

It is clear from (3.8) that motion on parts of the cycle Z_ε near the arcs P_2S_1 and P_1S_2 of Z_0 occurs with *finite* velocity. However, the parts close to S_1P_1 and S_2P_2 are traversed almost *instantaneously*, because on these parts the horizontal component of the phase-velocity vector is of order $1/\varepsilon$. In other words, in motion on Z_ε, a relatively slow smoothly varying state of the system alternates with a very

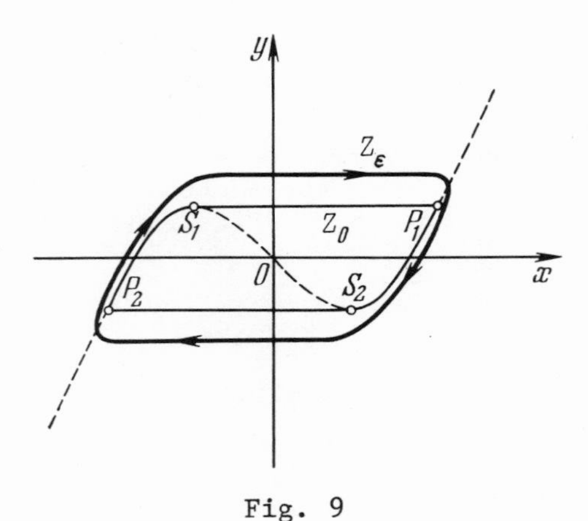

Fig. 9

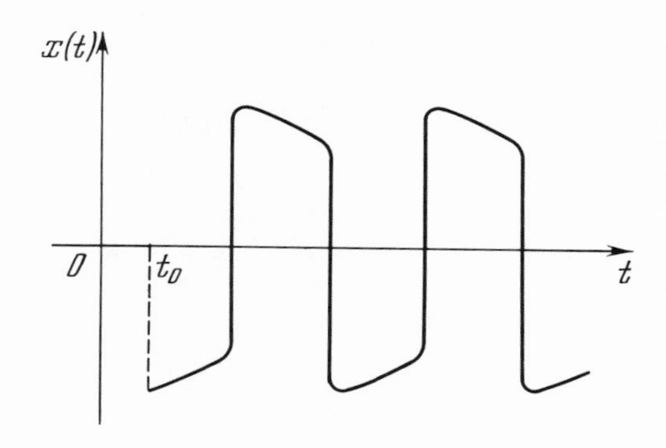

Fig. 10

rapid, almost discontinuous variation (Fig. 10). Periodic motion of this type is called *relaxation* (or discontinuous) *oscillation* [3, 7, 55].

This describes the general properties of the phase portrait of the Van der Pol equation. It consists of *rapid* and *slow* motions, with a transition or *surge* between these motions (near the points S_1 and S_2) and a *drop off* (close to P_1 and P_2).

This behavior also occurs in the phase portrait of any second-order system of the type (4.1). This property of trajectories of the system (4.1) close to certain parts of the curve Γ in the (x,y) plane with the equation

$$f(x, y) = 0 \tag{4.5}$$

can be explained kinematically in terms of the stability and instability of equilibrium positions of an auxiliary first-order equation. For example, a surge is due to a bifurcation of an equilibrium position.

The set of points on Γ where

$$\frac{\partial}{\partial x} f(x, y) < 0 \tag{4.6}$$

is called the *stable part* of Γ; the set of points on Γ where

$$\frac{\partial}{\partial x} f(x, y) > 0 \tag{4.7}$$

is called the *unstable part* of Γ. Stable and unstable parts are separated by points where

$$\frac{\partial}{\partial x} f(x, y) = 0; \tag{4.8}$$

for simplicity we assume that these are isolated points on Γ. For example, for the Van der Pol equation, the curve Γ [the cubic parabola (4.4)] consists of two stable parts $(-\infty, S_1)$ and $(+\infty, S_2)$ and one unstable part (S_1, S_2), separated by the two points S_1 and S_2 (Fig. 6).

We now consider the first of Eqs. (4.1),

$$\varepsilon \dot{x} = f(x, y); \tag{4.9}$$

we consider y in this equation as a parameter. For a fixed value of this parameter, for example, for $y = y_1$, Eq. (4.9) can have equilibrium positions among its solutions, say x_1. The definition of an equilibrium position implies that $f(x_1, y_1) = 0$; hence (x_1, y_1) is on the curve Γ. Conversely, if (x_1, y_1) is a point on Γ, then x_1 is an equilibrium position of (4.9) for $y = y_1$.

We thus have a one-to-one relation between points of Γ and all equilibrium positions of the family of equations (4.9) for all values of y. By virtue of (4.6) and (4.7), the stable parts of Γ consist of stable equilibrium positions, and the unstable parts consist of unstable equilibrium positions. The points separating these parts, i.e., the points of the curve (4.5) at which inequality (4.8) holds, are junctions between stable and unstable equilibrium positions. For example, in Fig. 11 the coordinate $y = y_2$ of the junction S is a bifurcation value of the parameter y; for $y < y_2$ in the

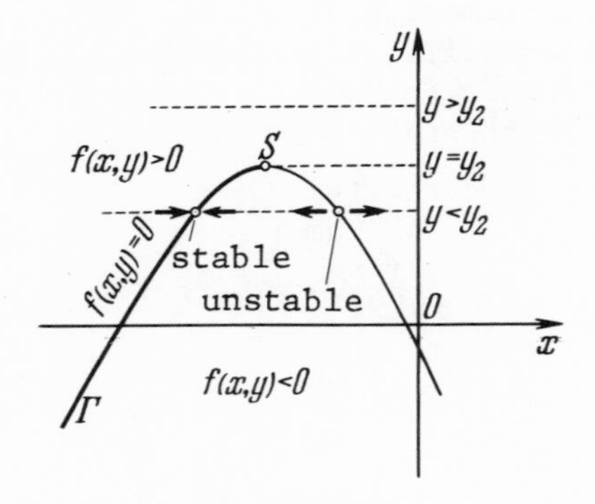

Fig. 11

neighborhood of S, Eq.(4.9) has two equilibrium positions (one stable, one unstable), and for $y > y_2$ close to S there is no equilibrium position.

By virtue of these considerations, we can treat motion of any phase trajectory of (4.1) as follows (Fig. 12). At each point (x,y) of the phase plane [or of the region in which (4.1) is being investigated], consider the phase-plane vector

$$v(x, y)=\left(\frac{1}{\varepsilon} f(x, y), g(x, y)\right)$$

of the system (4.1) (Fig. 7), and let $Q_0(x_1,y_1)$ be the initial point of the motion. If this point is at a finite distance from the curve (4.5), then the phase-velocity vector at this point, for a finite second component, has an infinite first component. Hence there is a rapid, almost instantaneous, change in the coordinate x for a very slight variation in y, i.e., motion on the trajectory of (4.1) will be close to motion on the line $y = y_1$, by virtue of the equation

$$\varepsilon x=f(x, y_1). \tag{4.10}$$

The nature of this motion does not change until the components of the phase-velocity vector become comparable

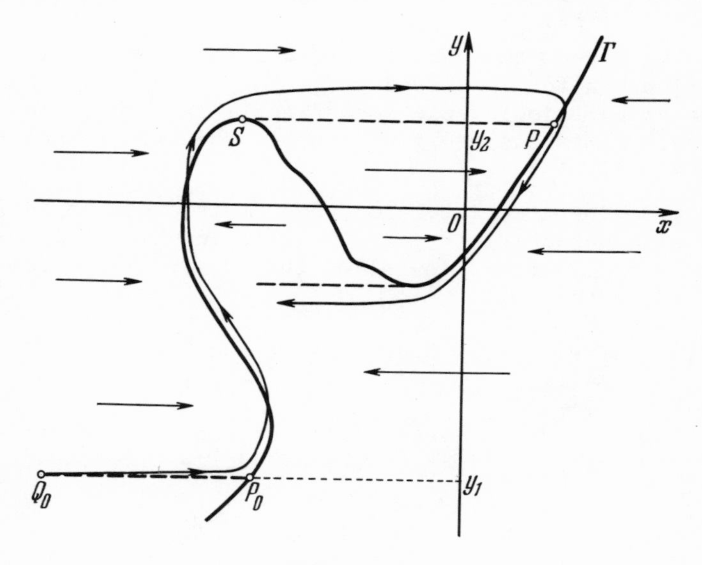

Fig. 12

in magnitude, i.e., until the phase point of (4.1) approaches the curve Γ (at a distance of order ε) or, equivalently, until x, according to (4.10), approaches one of the stable equilibrium positions . [If (4.10) has no stable equilibrium positions, a phase point of (4.1) having a high velocity goes to infinity, almost following the line $y = y_1$.]

The next part of the motion on a trajectory of (4.1) will be smooth and close to a stable part of Γ; it will be like motion on a stable equilibrium position of (4.9) for varying y. The variation of y is slow by virtue of the degenerate system (4.2).

If there are no equilibrium positions of (4.2) on the stable part of Γ, then it can happen that y, after a finite time, reaches a bifurcation value, for example $y = y_2$. [If there is no bifurcation value, the phase point of (4.1) slowly goes to infinity, remaining close to Γ.] For this value of y there is no accompanying stable equilibrium position, and the phase point of (4.1) rapidly moves (almost on the line $y = y_2$) into the neighborhood of the other stable equilibrium position of the equation

$$\varepsilon\dot{x} = f(x, y_2)$$

(or, if there is no other stable equilibrium position, it goes to infinity), etc.

It can happen that successive alternations of slow and fast motion can form a closed trajectory (Fig. 9). Then the corresponding periodic solution of (4.1) is a *relaxation oscillation*. It is easy to give examples of such closed trajectories (Fig. 13).

In the kinematic interpretation, x is called a *fast* and y a *slow* variable. Equation (4.9), in which y is considered as a parameter, is called the *fast-motion equation* corresponding to (4.1).

5. Systems of Arbitrary Order. Fast and Slow Motion. Relaxation Oscillations

We now consider a system (3.1) of arbitrary order with a small parameter multiplying certain derivatives; putting

$$x = (x^1, \ldots, x^k), \qquad y = (y^1, \ldots, y^l),$$
$$f = (f^1, \ldots, f^k), \qquad g = (g^1, \ldots, g^l),$$

we use the vector form of the equations:

$$\begin{cases} \varepsilon \dot{x} = f(x, y), \\ \dot{y} = g(x, y). \end{cases} \tag{5.1}$$

The phase space R^n, $n = k + l$, of the system (5.1) decomposes naturally into a direct sum of the k-dimensional space X^k and the l-dimensional space Y^l.

Let

$$\begin{cases} f(x, y) = 0, \\ \dot{y} = g(x, y) \end{cases} \tag{5.2}$$

be the corresponding *degenerate system*, i.e., the system obtained from (5.1) by putting $\varepsilon = 0$. The first equation

$$f(x, y) = 0 \tag{5.3}$$

in (5.2) describes an l-dimensional surface Γ in R^n; hence all trajectories of (5.2) are on this surface. Thus (5.2) does not have solutions with arbitrary initial points (x_0, y_0), but only solutions with initial points on Γ.

The question of the proximity of solutions of (5.1) and (5.2) has a meaning only for solutions of (5.1) starting in

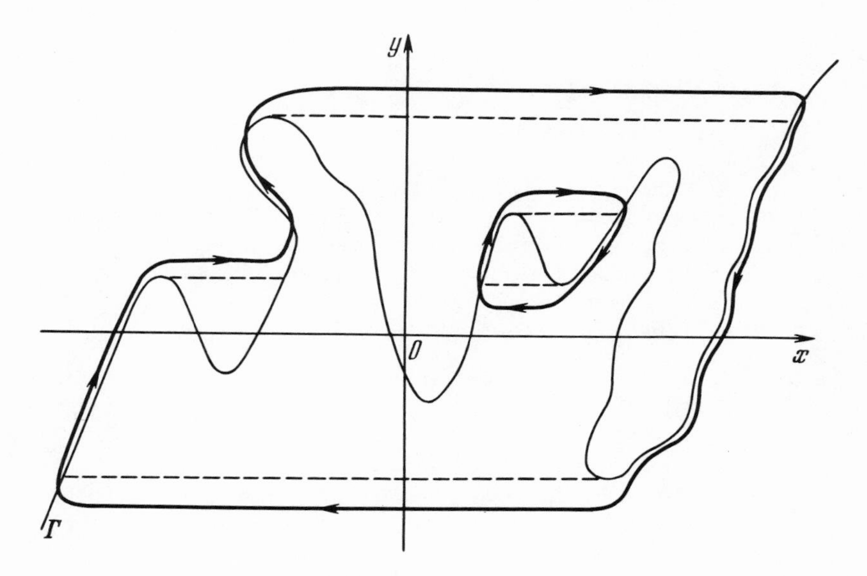

Fig. 13

a neighborhood of Γ that is small when ε is small. It turns out, however, that even these solutions of (5.1) do not always approach solutions of (5.2) when $\varepsilon \to 0$. It is also important to clarify under what conditions an arbitrary trajectory of (5.1), from its initial point (x_0, y_0) at a finite distance from the surface (5.3), is in a small (together with ε) neighborhood of this surface, the time passed on this section, etc.

All such questions will be considered in Chapters IV and V. Here we only give a qualitative description of the phase portrait of the system (5.1), and, using an example, we indicate the possibility of the existence of relaxation oscillations for a system of the type (5.1).

At each point (x,y) of the phase space R^n [or at each point of the region in which we are considering the system (5.1)], the system (5.1) determines the phase-velocity vector

$$v(x, y) = \left(\frac{1}{\varepsilon} f(x, y), \quad g(x, y)\right). \tag{5.4}$$

The second component of this vector has, as we shall see, a finite value, while the first is, in general, infinite. Hence it is characteristic of (5.1) to have *fast* and *slow* motion. Fast motion takes place *far* from the surface Γ and is almost parallel to the subspace X^k, while the slow motion is *near* Γ, where the first and second components of the vector (5.4) have similar magnitudes.

We shall now describe the general features of motion on an arbitrary phase trajectory of the system (5.1) (Figs. 14 and 15; see also Fig. 12). If the initial point $Q_0(x_1, y_1)$ is at a finite distance from the surface (5.3), we assume that y hardly varies, while x experiences a rapid, almost instantaneous, variation, i.e., motion on the trajectory of (5.1) differs only slightly from motion on the plane $y = y_1$ by virtue of the system

$$\varepsilon \dot{x} = f(x, y_1). \tag{5.5}$$

The system

$$\varepsilon \dot{x} = f(x, y), \tag{5.6}$$

with y considered as a (vector-valued) parameter, will be called the *rapid-motion equation system* corresponding to (5.1), and x will be called a *fast variable*.

We assume that, for each value of y, stationary solutions of (5.6) are equilibrium positions. Then, if (5.5)

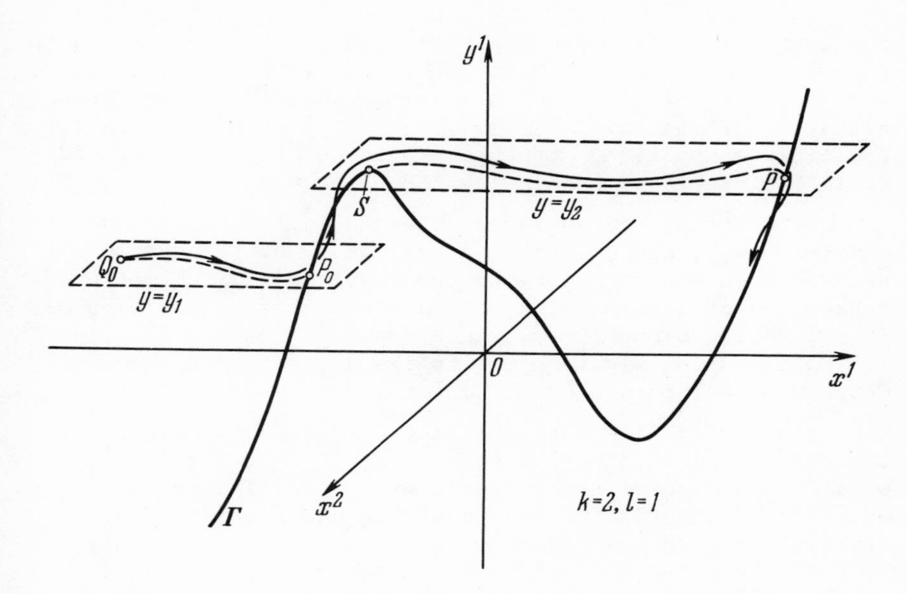

y^1
S
P
$y = y_2$
Q_0
P_0
$y = y_1$
0
x^1
x^2
Γ
$k=2, l=1$

Fig. 14

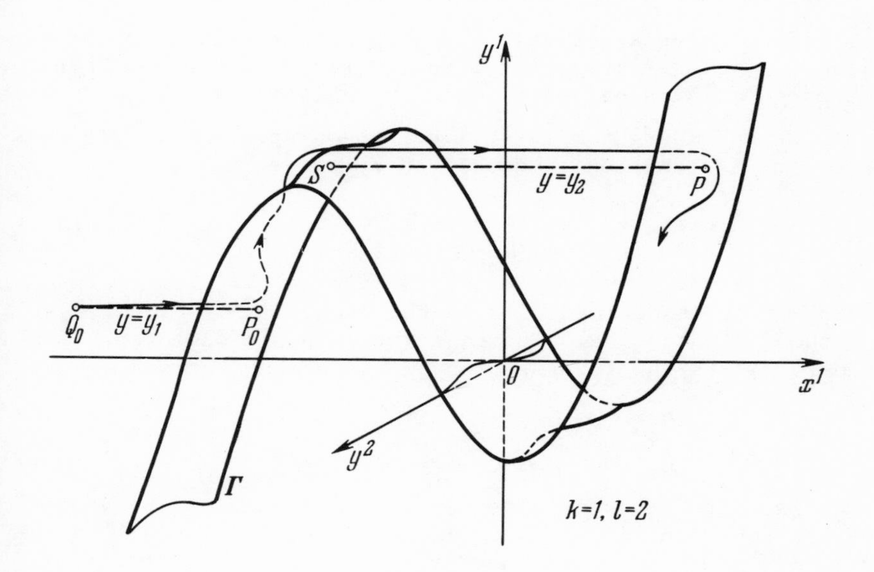

y^1
S
$y = y_2$
P
Q_0
$y = y_1$
P_0
0
x^1
y^2
Γ
$k=1, l=2$

Fig. 15

has a stable equilibrium position, the point x, according to (5.5), rapidly approaches one of them, say $x = x_2$; hence the phase point of (5.1) reaches a neighborhood (of order of magnitude ε) of the point $P_0(x_2, y_1)$. [If (5.5) has no stable equilibrium positions, the phase point of (5.1) goes rapidly to infinity, closely following the plane $y = y_1$.]

After this, the variables x and y in (5.1) will vary at similar rates, and motion on a trajectory of the system will proceed smoothly and close to the surface Γ. Such motion behaves as if accompanying a stable equilibrium position of (5.6), moving along the surface Γ as y varies. The variation of y proceeds slowly, subject to the degenerate system (5.2). Hence y is called a *slow variable*.

The motion of the phase point of (5.1) close to the surface Γ is maintained up to a bifurcation value $y = y_2$, when the associated stable equilibrium position ceases to exist [at a junction with an unstable equilibrium position of (5.6)]. Then the phase point of (5.1) moves suddenly (almost on the plane $y = y_2$) to the neighborhood of another stable equilibrium position of the system

$$\varepsilon\dot{x} = f(x, y_2)$$

(or goes to infinity, if there is no other such position), etc.

Successive alternation between slow and fast motion can lead to a closed trajectory; the corresponding periodic solution of (5.1) is called *relaxation oscillation*.

We consider Example 4, i.e., the system (3.9). In this case the fast motion is described by the equations

$$\begin{cases} \varepsilon\dot{x}^1 = -\alpha(y^1 - y^2) + \varphi(x^1) - x^2, \\ \varepsilon\dot{x}^2 = \alpha(y^1 - y^2) + \varphi(x^2) - x^1; \end{cases} \tag{5.7}$$

this motion is defined only in the square $|x^1| < 1$, $|x^2| < 1$ of the (x^1, x^2) plane. The equilibrium positions of the system (5.7) are the intersection points of the curves

$$K_1(x^1, x^2) \equiv -\alpha(y^1 - y^2) + \varphi(x^1) - x^2 = 0, \tag{5.8}$$

$$K_2(x^1, x^2) \equiv \alpha(y^1 - y^2) + \varphi(x^2) - x^1 = 0, \tag{5.9}$$

that is, the curves obtained by a translation of the graphs of the functions $x^2 = \varphi(x^1)$ and $x^1 = \varphi(x^2)$ (Fig. 5) in the directions of the coordinate axes. It is easily seen that, depending on the values of y^1 and y^2, the following situa-

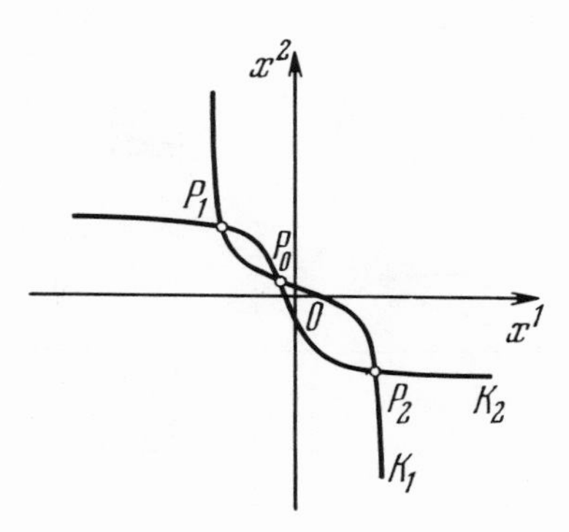

Fig. 16

tions can arise:

(a) the system (5.7) has three equilibrium positions P_1, P_2, and P_0, with P_1 and P_2 stable nodes and P_0 a saddle point (Fig. 16);

(b) the system (5.7) has two equilibrium positions $P_1 = P_0$ and P_2 (or $P_2 = P_0$ and P_1), the first being an unstable saddle-point node and the second being a stable node (Fig. 17);

(c) the system (5.7) has only one equilibrium position P_2 (or P_1), which is a stable node (Fig. 18).

The phase portraits for systems (5.7) corresponding to these cases are shown schematically in Figs. 19-21. It can easily be proved (for example by applying Bendixson's criteria

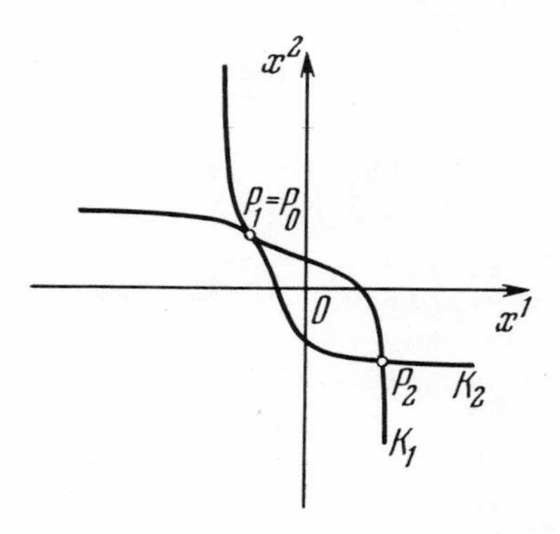

Fig. 17

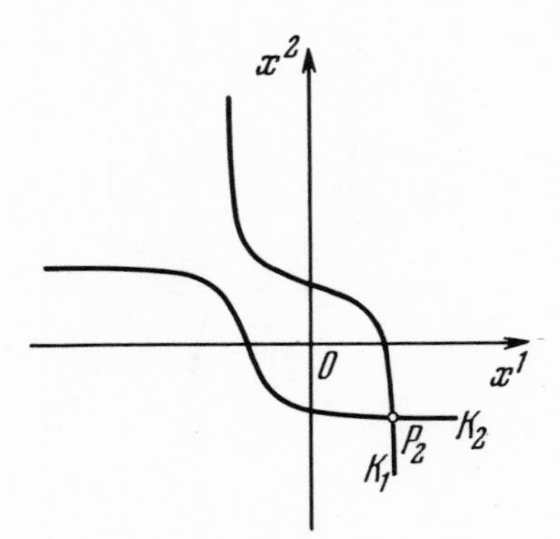

Fig. 18

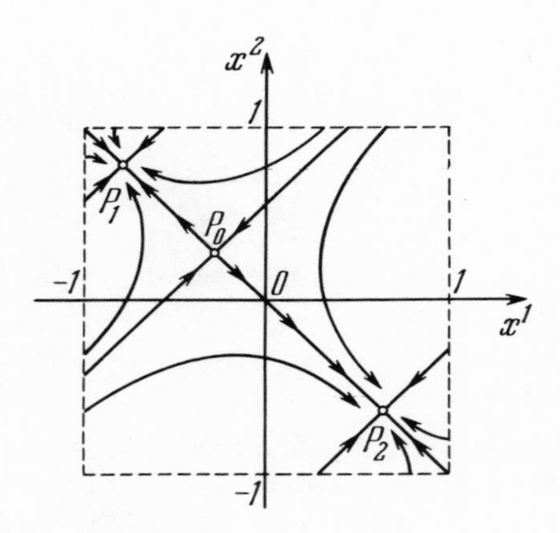

Fig. 19

[3, 30]) that (5.7) does not have closed trajectories for any values of the parameters y^1, y^2.

Now let $Q_0(x_1^1, x_1^2, y_1^1, y_1^2)$ be an arbitrary initial point of a trajectory of the system (3.9); we assume that, for $y = y_1 = (y_1^1, y_1^2)$, we have case (a) for the system (5.7) in the (x^1, x^2) plane (see Figs. 16 and 19; the other situations can be analyzed similarly). Since $x = (x^1, x^2)$ changes rapidly while $y = (y^1, y^2)$ changes very little, we assume that the trajectory of (3.9) starting at Q_0 remains close to the plane $y^1 = y_1^1$, $y^2 = y_1^2$, and rapidly approaches one of the stable

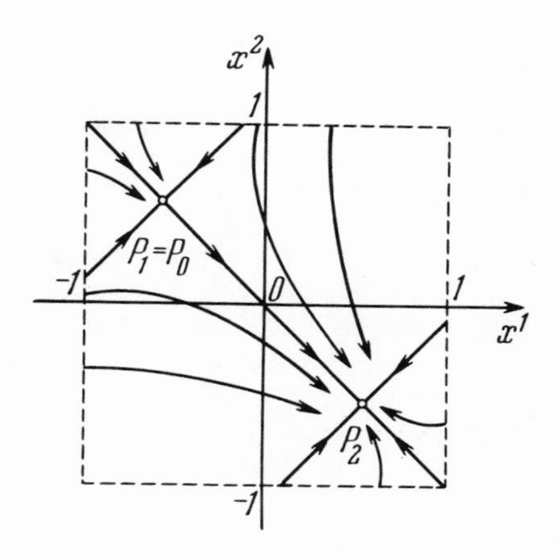

Fig. 20

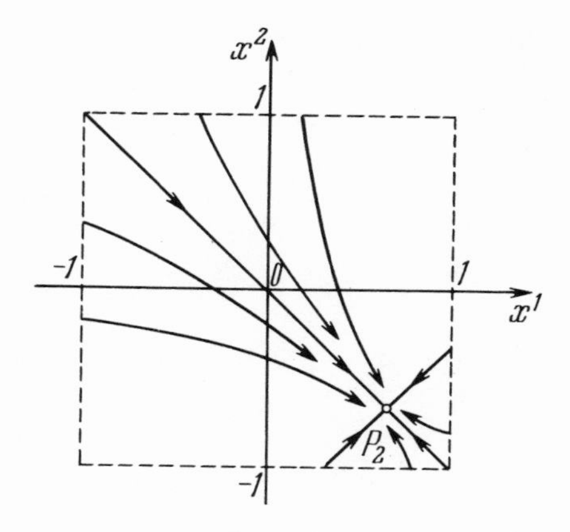

Fig. 21

equilibrium positions of the system

$$\begin{cases} \varepsilon\dot{x}^1 = -\alpha(y_1^1 - y_1^2) + \varphi(x^1) - x^2, \\ \varepsilon\dot{x}^2 = \alpha(y_1^1 - y_1^2) + \varphi(x^2) - x^1 \end{cases}$$

[see (5.5) and Fig. 14], for example the stable node $P_1(-a, a)$, where $a > 0$.

The variables x^1, x^2, y^1, y^2 then vary at similar rates, and the phase point of the system (3.9) is associated with the equilibrium position P_1, and moves slowly as y^1 and y^2

vary (cf. Fig. 15). This motion is easily traced. Since, by virtue of the last two equations in (3.9),

$$\dot{y}^1 = x^1, \quad \dot{y}^2 = x^2, \tag{5.10}$$

and $x^1 < 0$ and $x^2 > 0$ close to P_1, the difference $y^1 - y^2$ decreases when t increases. This means that the curve (5.8) is displaced upwards along the x^2 axis, and the curve (5.9) is displaced to the left in the direction of the x^1 axis. But then the stable equilibrium position P_1 moves clockwise relative to the saddle point P_0 and, at a certain time, for the bifurcation value $y = y_2 = (y_2^1, y_2^2)$, there is a junction and we no longer have stable equilibrium (Figs. 16, 17, and 19, 20).

The variables x^1, x^2 now begin to vary rapidly while the variables y^1, y^2 change very little. Hence we assume that the phase point of (3.9), remaining close to the plane $y^1 = y_2^1$, $y^2 = y_2^2$, moves rapidly to a stable-equilibrium position of the system

$$\begin{cases} \varepsilon\dot{x}^1 = -\alpha(y_2^1 - y_2^2) + \varphi(x^1) - x^2, \\ \varepsilon\dot{x}^2 = \alpha(y_2^1 - y_2^2) + \varphi(x^2) - x^1, \end{cases}$$

i.e., to the stable node $P_2(b, -b)$, where $b > 0$ (Figs. 18 and 21).

Here the variables x^1, x^2, y^1, y^2 again begin by varying at similar rates. We conclude from Eqs. (5.10) that the difference $y^1 - y^2$ is an increasing function of t, so that the curve (5.8) is displaced downwards in the direction of the x^1 axis while the curve (5.9) is displaced to the right in the direction of the x^2 axis. Thus, during a certain time interval, we successively have the situations shown in Figs. 17, 16, and 22, and finally Fig. 23. The phase point of (3.9) then moves rapidly to the remaining stable equilibrium position P_1, and the process is repeated.

Hence (3.9) has a periodic solution describing relaxation oscillations.

6. Solutions of the Degenerate Equation System

Here we examine in more detail the system (5.2) corresponding to (5.1) [or to (3.1)]. The first equation in (5.2) implies that trajectories of the degenerate system are on the l-dimensional space Γ described in R^n by Eq. (5.3) or,

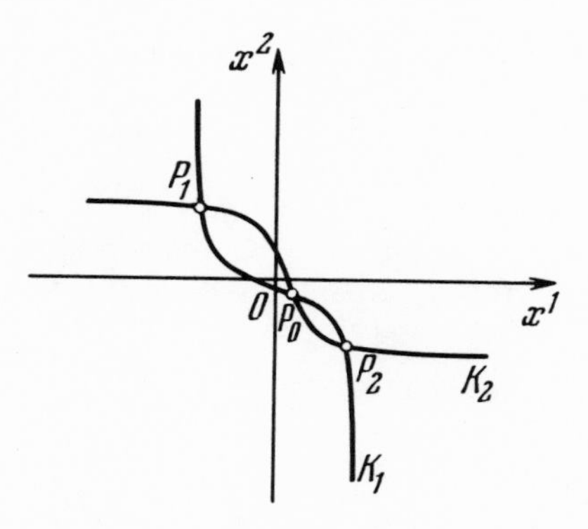

Fig. 22

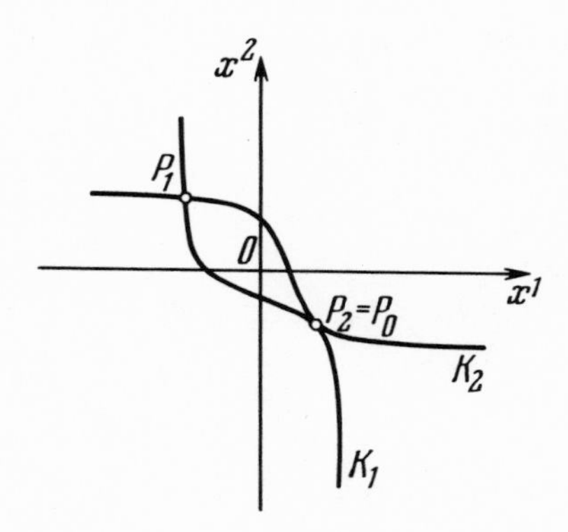

Fig. 23

in scalar form, the equations

$$\begin{cases} f^1(x^1, \ldots, x^k, y^1, \ldots, y^l) = 0, \\ \cdots\cdots\cdots\cdots\cdots \\ f^k(x^1, \ldots, x^k, y^1, \ldots, y^l) = 0. \end{cases} \tag{6.1}$$

The set of points on Γ at which all eigenvalues of the matrix

$$\mathfrak{A}(x^1, \ldots, x^k, y^1, \ldots, y^l) = \left\| \frac{\partial}{\partial x^\beta} f^\alpha(x^1, \ldots, x^k, y^1, \ldots, y^l) \right\|, \tag{6.2}$$

$$\alpha, \beta = 1, \ldots, k,$$

have negative real parts will be called the *stable region*

of Γ and will be denoted by Γ_-. The set of points of Γ at which

$$\det \mathfrak{A}(x^1, \ldots, x^k, y^1, \ldots, y^l) = 0 \tag{6.3}$$

will be called the *junction line* and will be denoted by Γ_0. The junction line Γ_0 is an $(l-1)$-dimensional subset of Γ and in general divides Γ into two or more parts.

The fast-motion equations (5.6) corresponding to the system (5.1) have the scalar form

$$\begin{cases} \varepsilon \dot{x}^1 = f^1(x^1, \ldots, x^k, y^1, \ldots, y^l), \\ \cdots\cdots\cdots\cdots\cdots \\ \varepsilon \dot{x}^k = f^k(x^1, \ldots, x^k, y^1, \ldots, y^l). \end{cases} \tag{6.4}$$

We recall that, in (6.4), $y^1, \ldots, y^l$ are considered as parameters. For fixed values

$$y^1 = y_0^1, \ldots, y^l = y_0^l,$$

of these parameters, the system (6.4) can have solutions that are equilibrium positions; let $(x_0^1, \ldots, x_0^k)$ be one of them. Then the point P_0,

$$(x_0, y_0) = (x_0^1, \ldots, x_0^k, y_0^1, \ldots, y_0^l), \tag{6.5}$$

by virtue of the definition of an equilibrium position, is on Γ (Fig. 24). Conversely, if (6.5) is a point on Γ, then

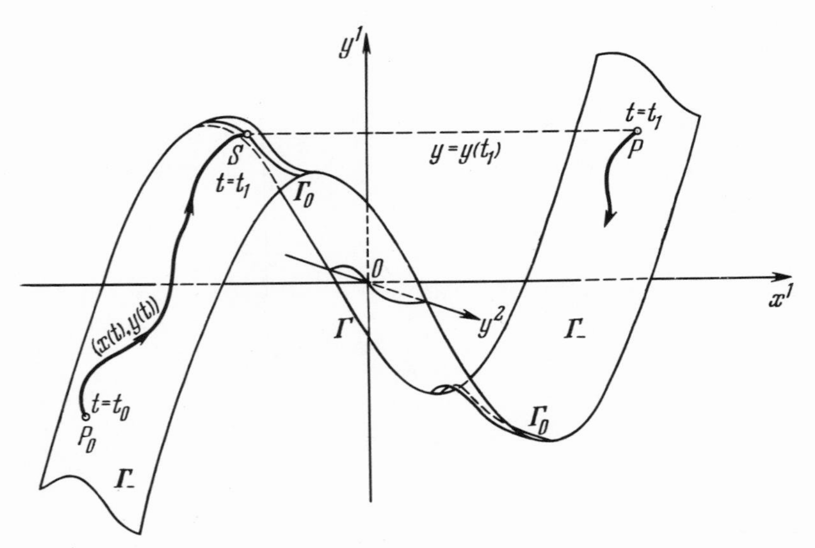

Fig. 24

$(x_0^1,\ldots,x_0^k)$ is an equilibrium position of the system (6.4) with $y^1,\ldots,y^l$ replaced by $y_0^1,\ldots,y_0^l$.

Hence we have a one-to-one relation between points of Γ and equilibrium positions of the family of systems (6.4) for all possible $y^1,\ldots,y^l$. It follows from our definition of a stable region Γ_- that Γ consists of stable equilibrium positions.

Now consider the degenerate system (5.2). To find its solutions it is clearly sufficient to solve relations (6.1) for $x^1,\ldots,x^k$,

$$x^i = \psi^i(y^1, \ldots, y^l), \quad i = 1, \ldots, k, \tag{6.6}$$

and then solve the following normal system of l equations with l unknowns [see the second of Eqs. (5.2)]:

$$\dot{y}^j = g^j(\psi^1(y^1,\ldots,y^l), \ldots, \psi^k(y^1,\ldots,y^l), y^1,\ldots,y^l), \qquad j = 1, \ldots, l. \tag{6.7}$$

Suppose that (6.5) is a point on Γ such that

$$x_0^i = \psi^i(y_0^1, \ldots, y_0^l), \quad i = 1, \ldots, k$$

[see (6.6)], and let

$$y^j = y^j(t), \quad j = 1, \ldots, l,$$

be the solution of the system (6.7) satisfying the initial conditions

$$y_0^j = y^j(t_0), \quad j = 1, \ldots, l.$$

It is clear that the system

$$\begin{aligned} &x^i = x^i(t) \equiv \psi^i(y^1(t), \ldots, y^l(t)), \quad i = 1, \ldots, k, \\ &y^j = y^j(t), \quad j = 1, \ldots, l, \end{aligned} \tag{6.8}$$

is the solution of the degenerate system (5.2) with initial value (6.5) for $t = t_0$. In other words, the trajectory of the solution (6.8) starts from the point (6.5) at time $t = t_0$ and lies on the surface Γ (Fig. 24).

It is easily seen that the solution (6.8) of the degenerate system (5.2) can be constructed only for values of $t \geq t_0$ for which the equations

$$f^i(x^1, \ldots, x^k, y^1(t), \ldots, y^l(t)) = 0, \quad i = 1, \ldots, k,$$

have a unique solution for $x^1,\ldots,x^k$. It follows from known results concerning implicit functions that a solution (6.8) exists from $t = t_0$ until the time $t = t_1$, when the determinant

$$\det \mathfrak{A}(x^1(t), \ldots, x^k(t), y^1(t), \ldots, y^l(t)) \tag{6.9}$$

first vanishes or, equivalently [see (6.3)], when the trajectory of the solution (6.8) first reaches a junction line Γ_0 at some point

$$S(x^1(t_1), \ldots, x^k(t_1), y^1(t_1), \ldots, y^l(t_1)) \tag{6.10}$$

(Fig. 24). In the general case, no conclusion can be drawn from the degenerate system (5.2) concerning the subsequent behavior of the solution (6.8); no trajectory of the degenerate system leaves the point S with increasing time.

However, if trajectories of the degenerate system (5.2) are considered as limits of trajectories of (5.1) for $\varepsilon \to 0$, then, by virtue of considerations in Sec. 5, the solution (6.8) can be extended to times $t > t_1$.

Suppose that, for $t_0 \le t < t_1$, the trajectory of a solution (6.8) is in the stable region Γ_- of Γ, and for $t = t_1$ this trajectory reaches a point S on the junction line Γ_0. Since the determinant (6.9) vanishes at the point (6.10), at least one eigenvalue of $\mathfrak{A}$ [see (6.2)] vanishes at this point; we assume that only one eigenvalue vanishes, and all the other $k - 1$ eigenvalues have negative real parts.* We also assume that the k-dimensional plane in R^n with the equations

$$y^1 = y^1(t_1), \ldots, y^l = y^l(t_1), \tag{6.11}$$

contains a trajectory of (6.4) [in which the parameters $y^1, \ldots, y^l$ are replaced by their values (6.11)], approaching the point $(x^1(t_1), \ldots, x^k(t_1))$ when $t \to -\infty$ and approaching a stable equilibrium position $(x^1_*, \ldots, x^k_*)$ when $t \to +\infty$.

Then we assume that the phase point of the degenerate system (5.2) has an instantaneous jump for $t = t_1$ from the point S [see (6.10)] to the point

$$P(x^1_*, \ldots, x^k_*, y^1(t_1), \ldots, y^l(t_1)) \tag{6.12}$$

lying on the surface Γ (more precisely, in the stable region Γ_-; see Fig. 24). Further motion of the phase point proceeds on the trajectory of the degenerate system (5.2) start-

*We also assume that other conditions for nonsingularity are satisfied; these conditions will be introduced in Chapter IV.

ing at time $t = t_1$ from the point P (6.12). If, during this motion, the phase point again reaches the junction line Γ_0, there is an instantaneous jump, etc. In the above interpretation, S is called a *junction point* and P is called the *drop point* following the junction point.

The vector-valued function $(x^1(t),\ldots,x^k(t),\ y^1(t),\ldots,\ y^l(t))$ that we have obtained is discontinuous (in general, all its components are discontinuous for $t = t_1$) and satisfies (5.2) everywhere except at junction points. It is naturally called a *discontinuous solution* of the degenerate system. The *trajectory of a discontinuous solution* is a discontinuous curve in the phase space R^n and is formed of successive parts of two types:

(a) parts in the stable region Γ_- of Γ on which the phase point passes a finite time;

(b) parts between a junction point and the following drop point, in planes parallel to the subspace X^k, on which the phase point moves instantaneously.

If successive parts of these types form a closed trajectory, we say that the degenerate system has a *discontinuous periodic solution.*

The definition of a discontinuous solution is particularly clear when x and y are scalar variables, so that the degenerate system has the form (4.2). By virtue of considerations in Sec. 4 concerning the nature of trajectories of (4.1), the trajectory of a discontinuous solution of the degenerate system (4.2) is formed of parts of the curve Γ traversed in a finite time, and horizontal segments from junction points to drop points traversed instantaneously (Fig. 12). The closed curve Z_0 in Figs. 8 and 9 is the trajectory of a discontinuous periodic solution of the degenerate system (4.3).

7. Asymptotic Expansions of Solutions with Respect to a Parameter

The fundamentals of the general theory of asymptotic methods and their applications to mathematical and applied problems are described in [10, 17, 24, 26, 27, 65]. Here we recall briefly only the basic problem of asymptotic expansion, with respect to a parameter, of solutions of system of ordinary differential equations, and find asymptotic expansions for solutions with various degrees of accuracy.

Consider the normal autonomous system

$$\dot{x} = F(x, \varepsilon) \tag{7.1}$$

[see (1.1) and (1.2)], involving the small parameter $\varepsilon > 0$, and let

$$x = \varphi(t, \varepsilon), \quad t_0 \leqslant t \leqslant T, \tag{7.2}$$

be a solution determined by a supplementary condition for each value of the parameter. Since, in general, we cannot find an explicit expression for the solution (7.2) for a fixed $\varepsilon > 0$, and it is even less likely that we can find a formula for $\varphi(t, \varepsilon)$ as a function of two variables, the problem arises of obtaining approximate representations of the function (7.2), i.e., of obtaining *approximate solutions* of (7.1).

If, for a small fixed $\varepsilon = \varepsilon^*$, we can find a function $x = \tilde{\varphi}(t, \varepsilon^*)$ such that

$$\| \varphi(t, \varepsilon^*) - \tilde{\varphi}(t, \varepsilon^*) \| \leqslant \delta \tag{7.3}$$

on the interval $t_0 \leq t \leq T$, where δ is a given number (say $\delta = 0.0001$), the function $x = \tilde{\varphi}(t, \varepsilon^*)$ is called an *approximate solution* of (7.1) for $\varepsilon = \varepsilon^*$ on the interval $t_0 \leq t \leq T$ with a given accuracy. This definition of an approximate solution is primarily related to *numerical methods*. At present, numerical methods of solving ordinary differential equations with contemporary electronic computers enable us to find, with a high degree of accuracy, solutions of practically any system of the form (7.1) for an arbitrary fixed value of the parameter.

However, calculation of functions $x = \tilde{\varphi}(t, \varepsilon^*)$, even for extremely fine sets of discrete values of the parameters, does not yield answers to many questions concerning the qualitative structure of solutions of (7.1) for small values of ε. It is therefore important to clarify the nature of the dependence of a solution (7.2) on ε when $\varepsilon \to 0$, i.e., to discover the limit properties of the function $\varphi(t, \varepsilon)$ of two variables when the second argument tends to zero. This problem arises when we wish to study solutions of (7.1) "for sufficiently small values of the parameter ε."

An effective method of solving this type of problem is to apply *asymptotic methods*, leading to the construction of approximate solutions by a method different from that described above, i.e., to the determination of asymptotic approximations. In (7.3), the deviation of an approximate solution from the exact solution is completely determined by a constant, while in an asymptotic approximation it is determined by a quantity of some order of smallness with respect to ε for $\varepsilon \to 0$.

A natural idea, based on the ordinary Taylor series, is to try to approximate the function (7.2), for sufficiently small values of the parameter, by a polynomial in ε, or by expanding the function in a series of powers of ε:

$$\varphi(t, \varepsilon) = \varphi_0(t) + \varepsilon\varphi_1(t) + \ldots + \varepsilon^k\varphi_k(t) + \ldots. \tag{7.4}$$

Then a partial sum of the series in (7.4) can serve as an approximation for $\varphi(t,\varepsilon)$, which becomes more accurate as we use more terms of the series and as ε becomes smaller.

Under certain conditions (see Theorems 1, 2, and 3) this approach actually succeeds; relations (1.8)-(1.10) are obtained in this way. For example, (1.8) means that $\lim_{\varepsilon\to 0}\varphi(t,\varepsilon)$ exists uniformly on the interval $t_0 \le t \le T$, and is equal to the function $x = \varphi_0(t)$ obtained by integrating the system (7.1) for $\varepsilon = 0$. In other words, the function $x = \varphi_0(t)$ is a *zeroth approximation* of the solution (7.2) for sufficiently small values of $\varepsilon > 0$.

Relation (1.9) should be understood as follows: There is a number $\varepsilon_0 > 0$ and a constant $M > 0$ such that, for arbitrary ε, $0 \le \varepsilon \le \varepsilon_0$,

$$\| \varphi(t, \varepsilon) - \Phi_{m-1}(t, \varepsilon) \| \leqslant M\varepsilon^m \tag{7.5}$$

on the whole interval $t_0 \le t \le T$, where

$$\Phi_{m-1}(t, \varepsilon) = \varphi_0(t) + \varepsilon\varphi_1(t) + \ldots + \varepsilon^{m-1}\varphi_{m-1}(t).$$

In this case one says that the function $x = \Phi_{m-1}(t, \varepsilon)$ is an *asymptotic approximation* for the solution (7.2) when $\varepsilon \to 0$, with accuracy of order ε^m. Finally, relation (1.10) shows that inequality (7.5) holds for any positive integer m (the constant M depends on m); here the series in (1.10) is called an *asymptotic expansion* of the solution (7.2) for $\varepsilon \to 0$.

If the conditions of Theorems 1, 2, and 3 are not satisfied, formulas (1.8)-(1.10) are generally inapplicable. This is the situation when the system (7.1) has the form (3.1), i.e., when the right side of the system depends on the small parameter ε in a *singular* fashion. For small $\varepsilon > 0$, no solution of such a system can be approximated by partial sums of a series of integral powers of the parameter.

In this situation, instead of a polynomial in ε, we attempt to effectively [i.e., without solving (7.1) for $\varepsilon > 0$]

construct a function of more general form, yielding an approximate expression for the solution (7.2) for $\varepsilon \to 0$ when quantities of higher than a certain order in ε are discarded. Such a function is called an *asymptotic approximation* of the solution (7.2) for $\varepsilon \to 0$ with the indicated degree of accuracy. Instead of power series we use series in certain functions of ε, whose partial sums can serve as asymptotic approximations of the solution (7.2) of as high an order as desired. Such series are called *asymptotic expansions* of the solution (7.2) for $\varepsilon \to 0$.

The procedure for obtaining asymptotic approximations and asymptotic expansions of the solution (7.2) for $\varepsilon \to 0$ is as follows in the general case. We first pose the question: Has the solution (7.2) a definite limit when $\varepsilon \to 0$, i.e., is there a function $x = \varphi_0(t)$, $t_0 \leq t \leq T$, with a representation

$$\varphi(t, \varepsilon) = \varphi_0(t) + \Delta_1\varphi(t, \varepsilon), \quad t_0 \leqslant t \leqslant T, \tag{7.6}$$

where $\Delta_1 \varphi(t,\varepsilon)$ tends to zero when $\varepsilon \to 0$, uniformly on $t_0 \leq t \leq T$? If such a representation exists, we write

$$\varphi(t, \varepsilon) = \varphi_0(t) + o(1), \quad t_0 \leqslant t \leqslant T; \tag{7.7}$$

the function $x = \varphi_0(t)$ is called the *zeroth approximation* for the solution (7.2) for $\varepsilon \to 0$.

It is now natural to ask what is the rate at which $\Delta_1 \varphi(t,\varepsilon)$ tends to zero, i.e., what is the order of $\Delta_1 \varphi(t,\varepsilon)$ with respect to ε when $\varepsilon \to 0$. For example, it can happen that, when $\varepsilon \to 0$, this function tends to zero like $\varepsilon^{2/3}$, uniformly on the interval $t_0 \leq t \leq T$. Then (7.6) can be written as

$$\varphi(t, \varepsilon) = \varphi_0(t) + O(\varepsilon^{2/3}), \quad t_0 \leqslant t \leqslant T.$$

The exact meaning of this relation is as follows [see (7.5)]. There is a number $\varepsilon_0 > 0$ and a constant $M > 0$ such that, for arbitrary ε, $0 < \varepsilon < \varepsilon_0$,

$$\| \varphi(t, \varepsilon) - \varphi_0(t) \| \leqslant M\varepsilon^{2/3}$$

for $t_0 \leq t \leq T$.

If we succeed in finding the principal part of $\Delta_1 \varphi(t,\varepsilon)$ for $\varepsilon \to 0$, i.e., if we can write it as

$$\Delta_1\varphi(t, \varepsilon) = \varepsilon^{2/3}\varphi_1(t) + \Delta_2\varphi(t, \varepsilon), \quad t_0 \leqslant t \leqslant T,$$

where $\Delta_2 \varphi(t,\varepsilon)$ tends to zero when $\varepsilon \to 0$ faster than $\varepsilon^{2/3}$, then (7.6) implies

$$\varphi(t, \varepsilon) = \varphi_0(t) + \varepsilon^{2/3}\varphi_1(t) + \Delta_2\varphi(t, \varepsilon), \quad t_0 \leqslant t \leqslant T,$$

or, equivalently,

$$\varphi(t, \varepsilon) = \varphi_0(t) + \varepsilon^{2/3}\varphi_1(t) + o(\varepsilon^{2/3}), \quad t_0 \leqslant t \leqslant T. \tag{7.8}$$

The precise meaning of this relation is as follows: There is a number $\varepsilon_0 > 0$ and a function $\omega(\varepsilon)$ tending to zero when $\varepsilon \to 0$ such that, for arbitrary ε, $0 < \varepsilon < \varepsilon_0$,

$$\|\varphi(t, \varepsilon) - \varphi_0(t) - \varepsilon^{2/3}\varphi_1(t)\| \leqslant \varepsilon^{2/3}\omega(\varepsilon)$$

for $t_0 \leq t \leq T$.

It can occur that $\Delta_2 \varphi(t,\varepsilon)$ tends to zero when $\varepsilon \to 0$, say, like $\varepsilon \ln(1/\varepsilon)$; then (7.8) takes the more accurate form

$$\varphi(t, \varepsilon) = \varphi_0(t) + \varepsilon^{2/3}\varphi_1(t) + O\left(\varepsilon \ln \frac{1}{\varepsilon}\right), \quad t_0 \leqslant t \leqslant T.$$

Separating out the principal part of $\Delta_2 \varphi(t,\varepsilon)$ for $\varepsilon \to 0$, we obtain the more accurate approximation

$$\varphi(t, \varepsilon) = \varphi_0(t) + \varepsilon^{2/3}\varphi_1(t) + \varepsilon \ln \frac{1}{\varepsilon}\varphi_2(t) + o\left(\varepsilon \ln \frac{1}{\varepsilon}\right), \quad t_0 \leqslant t \leqslant T;$$

if we express the remainder more accurately in this formula, we obtain

$$\varphi(t, \varepsilon) = \varphi_0(t) + \varepsilon^{2/3}\varphi_1(t) + \varepsilon \ln \frac{1}{\varepsilon}\varphi_2(t) + O(\varepsilon), \quad t_0 \leqslant t \leqslant T,$$

etc. By continuing this procedure indefinitely, if possible, we obtain an asymptotic expansion of the solution (7.2) for $\varepsilon \to 0$; for example

$$\varphi(t, \varepsilon) = \varphi_0(t) + \varepsilon^{2/3}\varphi_1(t) + \varepsilon \ln \frac{1}{\varepsilon}\varphi_2(t) + \varepsilon\varphi_3(t) +$$

$$+ \varepsilon^{4/3}\varphi_4(t) + \varepsilon^{5/3}\varphi_5(t) + \varepsilon^2 \ln \frac{1}{\varepsilon}\varphi_6(t) + \varepsilon^2\varphi_7(t) + \ldots, \quad t_0 \leqslant t \leqslant T.$$

We stress that, in the general case, this asymptotic series is not necessarily convergent [see (1.10)].

We have, of course, presented only the main features of a very simple approach to the determination of asymptotic approximations. In specific cases various complications arise, and modifications are necessary. Thus it can happen that there is no representation (7.6), but

$$\varphi(t, \varepsilon) = \varphi_0(t, \varepsilon) + \Delta_1\varphi(t, \varepsilon), \quad t_0 \leqslant t \leqslant T, \tag{7.9}$$

where $\varphi_0(t,\varepsilon)$ can be effectively calculated and $\Delta_1\varphi(t,\varepsilon)$ tends to zero when $\varepsilon \to 0$ uniformly on $t_0 \leq t \leq T$. In this case it is useful to find the principal part of $\Delta_1\varphi(t,\varepsilon)$ for $\varepsilon \to 0$. If

$$\Delta_1\varphi(t, \varepsilon) = \varepsilon^{2/3}\varphi_1(t, \varepsilon) + \Delta_2\varphi(t, \varepsilon), \quad t_0 \leqslant t \leqslant T,$$

where $\Delta_2 \varphi(t, \varepsilon)$ tends to zero when $\varepsilon \to 0$ more rapidly than $\varepsilon^{2/3} \varphi_1(t,\varepsilon)$, then (7.9) yields

$$\varphi(t, \varepsilon) = \varphi_0(t, \varepsilon) + \varepsilon^{2/3} \varphi_1(t, \varepsilon) + \Delta_2 \varphi(t, \varepsilon), \quad t_0 \leqslant t \leqslant T,$$

and we can proceed similarly. Estimates must be made of the size of the remainder at each step.

A typical situation is that in which the solution (7.2) has different asymptotic expansions on different parts of the interval $t_0 \leq t \leq T$ under consideration; in this case no asymptotic expansion of the solution (7.2) exists on the whole interval $t_0 \leq t \leq T$. A distinct expansion for each part must be found, such that expansions are in agreement at junction points, and such that their combination yields an asymptotic approximation to (7.2) with the required accuracy.

We note that, for different components of the solution (7.2), the expression $o(1)$ in (7.7) can denote functions tending to zero at different rates when $\varepsilon \to 0$. Then the components of the solution (7.2) will have asymptotic approximations of different structures.

8. A Sketch of the Principal Results

Here we give a short description of the contents of the chapters to follow, and indicate certain other results concerning the problems discussed in Secs. 4 and 5. We first note that the physical and mathematical fundamentals of relaxation oscillations, and also many concrete examples of oscillating systems of this type, can be found in the monographs [3, 7, 25, 30, 55, 63].

Chapter II is devoted to a detailed study of the second-order system (4.1). Our objective is to find asymptotic formulas yielding approximations to trajectories of this system with arbitrarily high accuracy with respect to ε for $\varepsilon \to 0$. It turns out that a trajectory of (4.1) with a given initial condition can be separated in a special fashion into several parts, such that we can obtain an asymptotic representation with any desired accuracy on each part; the asymptotic approximations coincide at the boundary points between the parts. It should be stressed that the asymptotic expansions for the trajectory on some parts are obtained in nonnegative integral powers of ε, while on other parts they are obtained as series of terms $\varepsilon^{n/3} \ln^{\nu}(1/\varepsilon)$, where n and ν are nonnegative integers. The coefficients in all expansions are determined recurrently without the need for integration of (4.1), i.e., directly in terms of the functions $f(x,y)$ and $g(x,y)$.

In Chapter III we obtain conditions for (4.1) to have a stable periodic solution with the properties of a relaxation oscillation; these conditions are obtained in terms of $f(x,y)$ and $g(x,y)$. Results from Chapter II are used to carry out the asymptotic calculation, with any given degree of accuracy, of a limit cycle Z_ε of the system (4.1) corresponding to a relaxation oscillation. Of special interest is the calculation of one of the most important characteristics of a relaxation oscillation — its period. We find that the period T_ε of Z_ε has the asymptotic expansion

$$T_\varepsilon = K_{0,0} + \sum_{n=2}^{\infty} \varepsilon^{n/3} \sum_{\nu=0}^{\pi(n-2)} K_{n,\nu} \ln^\nu \left(\frac{1}{\varepsilon}\right),$$

where $\pi(n)$ is a periodic function defined for nonnegative integral values of its argument:

$$\pi(n) = \begin{cases} k, & \text{if} \quad n = 3k, \\ k+1, & \text{if} \quad n = 3k+1, \\ k, & \text{if} \quad n = 3k+2, \end{cases}$$

and the $K_{n,\nu}$ are effectively determined without integration of (4.1). As an example we consider the determination of an asymptotic approximation for the period of the relaxation oscillation for a system described by Van der Pol's equation (3.6), (3.8).

The results in Chapters II and III are obtained by using ideas due to L. S. Pontryagin [44, 43] concerning second-order systems, further developed in [35, 50]; see also [37, 52]. These results were published for the first time; some further results can be found in [48] and [51]. Of the articles devoted to the asymptotic theory of relaxation oscillations in second-order systems, we note [9, 19, 22, 26, 28, 59, 61, 62]. There is an asymptotic calculation of a limit cycle for Van der Pol's equation by A. A. Dorodnitsyn [18] (see also [7]) in which he obtains the first four terms of the expansion for the period of relaxation oscillations.

The properties of solutions of the multidimensional system (5.1) depend essentially on the stationary solutions of the fast-motion equation system (5.6). A simpler investigation can be carried out if we assume that only the equilibrium positions of the system (5.6), for all y, can be stationary solutions. This case is investigated in Chapter IV, where we derive asymptotic formulas for approximations to trajectories of the system (5.1) with accuracy up to order ε for $\varepsilon \to 0$. To find the coefficients in the asymptotic representations, we need integrate only the degenerate system (5.2). The problem of finding approximations for trajectories of the system (5.1) with any desired degree of accuracy has not yet been solved.

The multidimensional system (5.1) can also have periodic solutions with the properties of relaxation oscillations. Chapter V is devoted to the derivation of sufficient conditions for (5.1) to have at least one closed trajectory corresponding to a relaxation oscillation; these conditions are obtained in terms of the degenerate system (5.2). Results obtained in Chapter IV are used to find an asymptotic formula for the period of the relaxation oscillations, with quantities of order $\varepsilon^{2/3}$ and $\varepsilon \ln(1/\varepsilon)$ taken into account but with higher-order quantities neglected. No asymptotic expansion for the relaxation oscillation period has yet been obtained for the multidimensional case.

In Chapters IV and V we reproduce, with some extra material, the important developments [38, 43, 34] of L. S. Pontryagin's original ideas [44]; see also [37, 39, 40]. The essence of these results first appeared in monographs [36, 41, 42]. In Chapters IV and V we use methods and results due to A. N. Tikhonov [56, 57] and A. B. Vasil'eva [12, 13]; see also [14]. The asymptotic theory of relaxation oscillations in multidimensional systems is investigated in [29, 53].

If the condition that, for each value of y, the only stationary solutions of the system (5.6) are equilibrium positions does not hold, many interesting but difficult problems arise. Their investigation lies outside the scope of this book, and we only state several such problems and describe a few results. More details can be found in [8].

There is an interesting case in which each solution of the system of fast-motion equations (5.6) tends, for $t \to \infty$, to an exponentially stable periodic solution. We shall give the general properties of motion of the phase point of (5.1) in this case.

Making the substitution $t = \varepsilon\theta$, we write (5.6) as

$$\frac{dx}{d\theta} = f(x, y)$$

(y is a parameter), and denote its periodic solution $L(y)$ by

$x = \varphi(\theta,y)$ and its period by $T(y)$. We introduce the averaged system

$$\frac{d\bar{y}}{dt} = \bar{g}(\bar{y}), \quad \bar{g}(\bar{y}) \equiv \frac{1}{T(\bar{y})} \int_0^{T(\bar{y})} g(\varphi(\theta, \bar{y}), \bar{y})\, d\theta.$$

If the initial point $Q_0(x_1,y_1)$ is at a finite distance from the cycle $L(y_1)$, the phase point of (5.1) rapidly (during a time of order ε) approaches the cycle. The slow variable $y(t,\varepsilon)$ is then close to the averaged solution $\bar{y}(t)$, and the fast variable $x(t,\varepsilon)$ remains close to the cycles $L(\bar{y}(t))$ and performs rapid oscillations about them with a period in t differing only slightly from $\varepsilon T(\bar{y}(t))$. Here essentially different situations can arise, depending on whether the averaged system has, for a stationary solution, a stable equilibrium position or a periodic solution.

Precise statements of the results of investigating motion of the phase point of the system (5.1) in the above case are given in [44, 46, 47, 49]. The investigation of this and analogous problems is closely related to methods developed by N. N. Bogolyubov and Yu. A. Mitropol'skii [7, 27, 32, 33, 15].

Various other situation which can occur in the systems of fast-motion equations have been investigated: the system has a Hamiltonian or has first integrals [31, 4]; the system has a degenerate equilibrium position or a degenerate limit cycle [6]; the system is two-dimensional and has an unstable limit cycle, and the part of the plane in the interior of the limit cycle is an attraction region of an unstable focus [64], etc. However, in all these cases, the general nature of the behavior of trajectories of the system (5.1) has not yet been determined.

Finally, the system (5.1) has been investigated under the assumption that the degenerate system (5.2) has a closed trajectory L completely in the stable region Γ_- of the surface Γ. In this case, under certain conditions, (5.1) has a periodic solution whose trajectory is near the curve L [60, 5].

CHAPTER II

SECOND-ORDER SYSTEMS. ASYMPTOTIC CALCULATION OF SOLUTIONS

Investigations of many systems with one degree of freedom lead to normal second-order autonomous systems of ordinary differential equations with a small parameter multiplying certain derivatives. For these systems, we give detailed answers to the questions posed in Chapter I concerning the dependence of solutions on the parameter.

Asymptotic expansions are obtained for phase trajectories of second-order systems with a small parameter multiplying a derivative. The resulting formulas can be used to calculate approximately any finite part of a trajectory with any desired degree of accuracy.

1. Assumptions and Definitions

We shall study the phase trajectories of the system

$$\begin{cases} \varepsilon\dot{x}=f(x, y), \\ \dot{y}=g(x, y), \end{cases} \tag{1.1}$$

where x and y are scalar functions of the independent variable t and ε is a small positive parameter. The right-hand sides of (1.1) will be assumed to be defined in the whole (x, y) phase plane and to be sufficiently smooth, i.e., differentiable with respect to x and y the number of times required by the reasoning used.

The system (1.1), which we call a *nondegenerate system*, corresponds to the *degenerate system* obtained by putting ε equal to zero in (1.1):

$$\begin{cases} f(x, y)=0, \\ \dot{y}=g(x, y). \end{cases} \tag{1.2}$$

In our reasoning, an important part is played by the curve Γ defined in the (x,y) plane by the first of Eqs. (1.2),

$$f(x, y)=0. \tag{1.3}$$

We make several assumptions concerning the geometrical properties of this curve; we use standard terminology.

We assume that the curve Γ (in general not necessarily connected) consists of *ordinary points*, i.e., at each point of the curve (1.3)

$$[f'_x(x, y)]^2 + [f'_y(x, y)]^2 > 0. \qquad (1.4)$$

The points of Γ at which

$$f'_x(x, y) = 0 \qquad (1.5)$$

will be called *nonregular;* other points will be called *regular*. We assume that nonregular points on Γ are *isolated* (although there can be an infinite number of nonregular points if Γ is infinite), and that all these points are *nondegenerate,* in the sense that, at each of them,

$$f''_x(x, y) \neq 0. \qquad (1.6)$$

Parts of Γ formed of only regular points will be called *regular parts*. For example, Fig. 25 shows a curve Γ having six nonregular points dividing Γ into seven regular parts.

Consider any regular part of Γ between the nonregular points $S^*(s_1^*, s_2^*)$ and $S(s_1, s_2)$. Since $f'_x(x,y)$ does not vanish on this part of the curve, it follows from the implicit function theorem that Eq. (1.3) has a unique solution x, expressible in explicit form as

$$x = x_0(y), \qquad (1.7)$$

where the argument y takes values on the interval between s_2^* and s_2 (Fig. 25; the cases $s_2^* = -\infty$ and $s_2 = \infty$ are permitted).

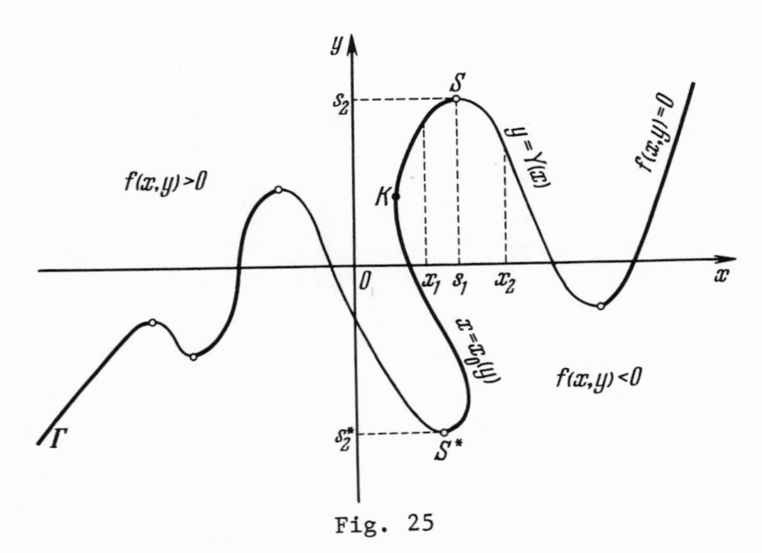

Fig. 25

Then

$$f(x_0(y), y) \equiv 0 \tag{1.8}$$

on the range of y between s_2^* and s_2. Differentiation of (1.7) yields

$$x_0'(y) = -\frac{f_y'(x_0(y), y)}{f_x'(x_0(y), y)}, \tag{1.9}$$

which determines the tangent to Γ at points of the regular part of Γ under consideration. A regular part can contain points at which the tangent is parallel to Oy (for example, the point K in Fig. 25); at these points $f_y'(x,y) = 0$.

Consider a nonregular point $S(s_1, s_2)$. In the neighborhood of this point, (1.3) does not have a unique solution x, because $f_x'(S) = 0$ and $f_y'(S) \neq 0$ [see (1.5) and (1.6)]. In other words, the nonregular point is a local extremal point on Γ. However condition (1.4) implies that $f_y'(S) \neq 0$; hence $f_y'(x,y)$ does not vanish in the neighborhood of S, and Eq. (1.3) of the curve Γ can be expressed, in some neighborhood of S, in the explicit form

$$y = Y(x),\ x_1 \leqslant x \leqslant x_2,\ s_1 \in (x_1, x_2) \tag{1.10}$$

(see Fig. 25). Thus $Y(s_1) = s_2$ and

$$f(x, Y(x)) \equiv 0,\ x_1 \leqslant x \leqslant x_2. \tag{1.11}$$

Differentiating this identity twice and using (1.5) and (1.6), we find that

$$Y'(s_1) = 0, \quad Y''(s_1) = -\frac{f_x''(S)}{f_y'(S)} \neq 0. \tag{1.12}$$

Hence Taylor's formula for the function (1.10) can be written as

$$y = s_2 - \frac{f_x''(S)}{2f_y'(S)}(x - s_1)^2 + \frac{Y'''(s_1 + \vartheta(x - s_1))}{6}(x - s_1)^3, \tag{1.13}$$

where $\vartheta \in [0, 1]$ is a constant and $x_1 \leqslant x \leqslant x_2$. Hence condition (1.6) for the nondegeneracy of a nonregular point is equivalent to the assumption that Γ, in a small neighborhood of this point, behaves roughly like a second-degree parabola in the vicinity of its vertex.

The regular part (1.7) between the nonregular points S^* and S of the curve Γ is called a *stable part* of the curve if, for all y between s_2^* and s_2,

$$f'_x(x_0(y),\ y) < 0. \tag{1.14}$$

Since the derivative $f'_x(x,y)$ does not vanish on a nonregular part, the function (1.7) is a nonmultiple root of Eq. (1.3); hence $f(x,y)$ has opposite signs on opposite sides of Γ. Inequality (1.14) implies that $f(x,y)$, on passage through the graph of a stable part of Γ in the direction of *increasing* x, changes sign from plus to minus. A regular part (1.7) on which $f'_x(x_0(y),\ y) > 0$, is called an *unstable part*. The heavy line in Fig. 25 indicates the stable parts of Γ.

Analysis of the phase-velocity field of the nondegenerate system (1.1) shows (see Chapter I, Sec. 4) that, if a regular part of Γ is stable, then, in this neighborhood, the vector field is directed towards the graph of this part; if it is unstable, the vector field is directed away from the graph of this part (Fig. 7). The assumption that nonregular points are nondegenerate implies that each of these points is a boundary between stable and unstable parts that meet at this point, in the fashion that two branches of a second-degree parabola meet at a vertex (Fig. 11).

We first assume that there are no equilibrium positions of (1.1) on stable parts, and also that no nonregular point is an equilibrium position of the system. In other words, we assume that $g(x,y) \neq 0$ at all points of the stable part and at all nonregular points of Γ.

A nonregular point S of Γ will be called a *junction point* if

$$\operatorname{sign}\,[f''_x(S)\, f'_y(S)\, g(S)] = 1 \tag{1.15}$$

at S. This relation and inequality (1.13) imply that, if a junction point S is a local maximum of Γ, then $g(S) > 0$, while if it is a local minimum of Γ, $g(S) < 0$. Thus, near a junction point of the type shown in Fig. 26, the phase-velocity vectors of a nondegenerate system (1.1) are directed upwards, while, near a junction point of the type in Fig. 27, the vectors are directed downwards. Hence a nonregular point is a junction point if, in its neighborhood, the phase-velocity vectors of (1.1) are in the direction of the convex side of Γ.

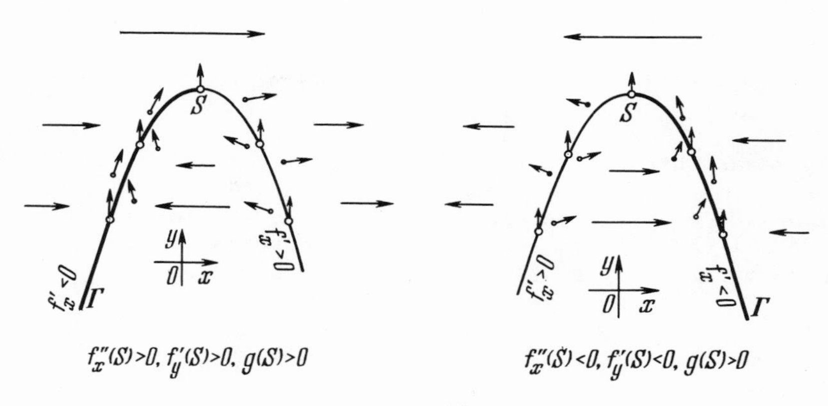

Fig. 26

We note that, on any stable part adjacent to a junction point S, $g(x,y)$ has the same sign as at S. If a stable part is bounded at both ends by nonregular points (that is, it does not go to infinity), one is a junction point and the other is not a junction point.

Let (x^*,y^*) be any point in the phase plane of the system (1.1). We say that this point is in the attraction region of a stable part (1.7) between nonregular points S^* and S if the solution $x = x(\theta)$ of the equation

$$\frac{dx}{d\theta}=f(x, y^*) \tag{1.16}$$

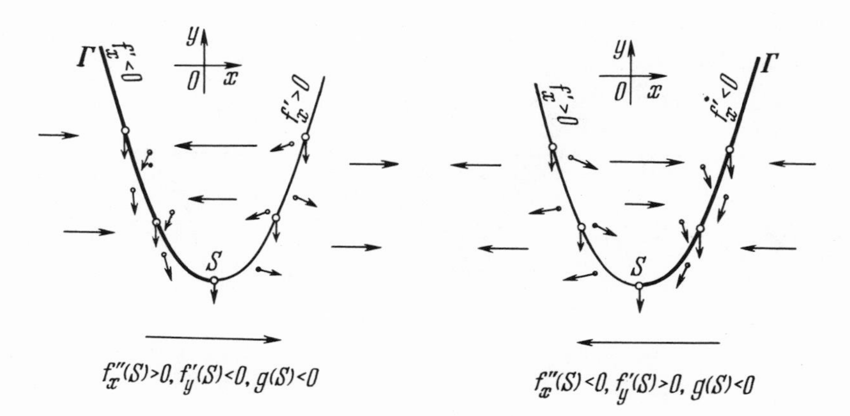

Fig. 27

with the initial condition $x(0) = x^*$ is such that

$$\lim_{\theta \to +\infty} x(\theta) = x_0(y^*). \tag{1.17}$$

This means that the attraction region of a stable part (1.7) consists of all points (x^*, y^*) for which y^* is in an interval of the y axis with ends at s_2^* and s_2, and for each fixed y^*, x^* can take values for points of the attraction region of the asymptotically stable equilibrium position $x = x_0(y^*)$ of Eq. (1.16), in which y^* is considered as a parameter, is called the *fast-motion equation* corresponding to (1.1); it is obtained from the first equation of (1.1) by changing to the *fast time* $\theta = t/\varepsilon$.

If a point (x^*, y^*) is in the attraction region of the stable part (1.7), then the intersection $P^*(x_0(y^*), y^*)$ of the line $y = y^*$ with the graph of this part is called the *drop point* from the point (x^*, y^*) (Fig. 28).

Finally, we assume that no two nonregular points of Γ have the same ordinates. Let $\tilde{S}(\tilde{s}_1, \tilde{s}_2)$ be a junction point satisfying the following condition: It is in the boundary of the attraction region of some stable part (1.7), but is distinct from the ends S^* and S (Fig. 28). This means that $\tilde{s}_2$ is in the interval between s_2^* and s_2, and the unstable

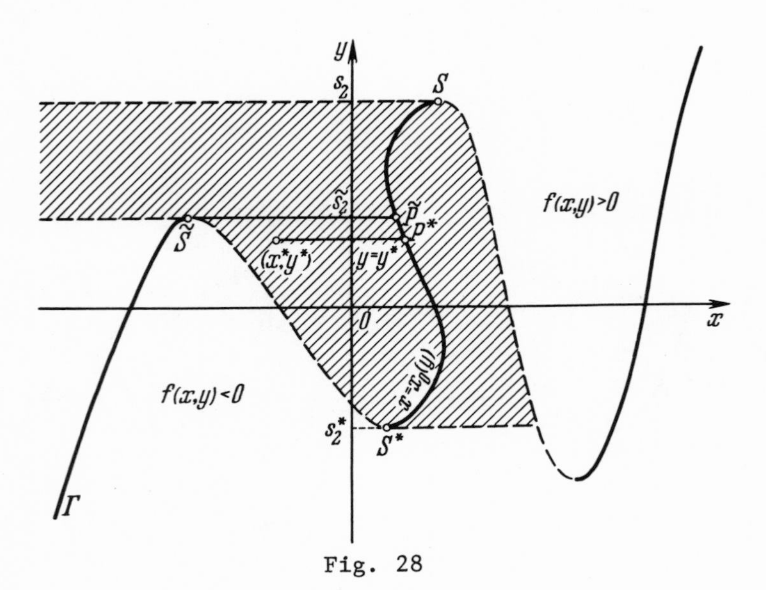

Fig. 28

direction from the semistable equilibrium position $x = \tilde{s}_1$ of Eq. (1.16) for $y^* = \tilde{s}_2$ has the property (1.17). Then the point $\tilde{P}(x_0(\tilde{s}_2), \tilde{s}_2)$ of intersection of the line $y = \tilde{s}_2$ with the graph of the stable part (1.7) will be called the *drop point following the junction point* $\tilde{S}$. If the junction point $\tilde{S}$ does not satisfy the foregoing condition, a drop point following the junction point is not defined.

2. The Zeroth Approximation

The idea of a *discontinuous solution* can be introduced for the system (1.2) (Chapter I, Sec. 6); it is then natural to understand phase trajectories of the system to be the trajectories of its discontinuous solutions. However, it can be seen that the phase portrait of the system (1.2) can also be described directly. Namely, *the phase trajectory of (1.2) with initial point* Q is the continuous curve $\mathfrak{T}$ in the (x,y) plane obtained by applying the following rules.

(1) If the point Q is not on Γ and is not in the attraction region of any stable part of this curve, the trajectory $\mathfrak{T}$ is a ray parallel to the Ox axis from Q with the direction of the vector $(f(Q),0)$. The phase point of (1.2) passes instantaneously to infinity on this way.

(2) If Q is not on Γ and is in an attraction region of a stable part of the curve, the first part of the trajectory $\mathfrak{T}$ is the segment parallel to the Ox axis from Q to the drop point of Q on Γ. The phase point of (1.2) traverses this segment instantaneously.

(3) If Q is in some stable part of Γ, and the boundary of this part contains a junction point, the first section of the trajectory $\mathfrak{T}$ is the arc of Γ from Q to the junction point. The phase point of (1.2) traverses this arc in a completely determined finite time. Such an arc of Γ is called a *slow part of* $\mathfrak{T}$.

(4) If Q is in some stable part of Γ and the boundary of this part does not contain a junction point, the trajectory $\mathfrak{T}$ is the arc of Γ from Q to infinity.

(5) If Q is a junction point of Γ and there is a drop point following this junction, the first section of the trajectory $\mathfrak{T}$ is the segment parallel to Ox from Q to the drop point. The phase point of (1.2) traverses this segment instantaneously. This type of segment is called a *fast-motion part* of $\mathfrak{T}$.

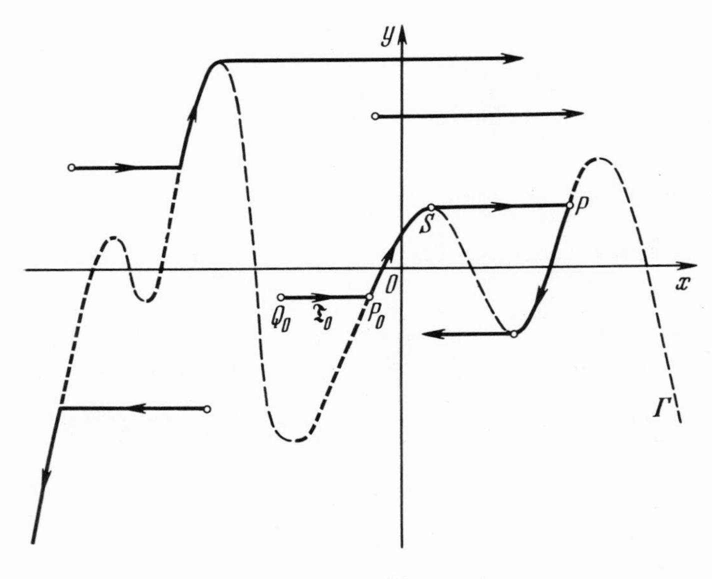

Fig. 29

(6) If Q is a junction point of Γ, and no drop point follows this junction, the trajectory $\mathfrak{T}$ is a ray parallel to the Ox axis from Q in the direction of the vector $(f''_x(Q), 0)$.

Figure 29 shows phase trajectories of (1.2) drawn according to these rules. These phase trajectories of the degenerate system, beginning at points of the unstable parts of Γ or at nonregular points of this curve that are not junction points, have not been found. We stress that the phase trajectory of the system (1.2) can be effectively traced if the functions $f(x,y)$ and $g(x,y)$ and the initial point are known.

Now let $Q_0(x_0,y_0)$ be a fixed point of the (x,y) phase plane not on an unstable part of Γ and not a nonregular point. We write

$$x = x(t, \varepsilon), \quad y = y(t, \varepsilon) \tag{2.1}$$

for the solution of the nondegenerate system (1.1) satisfying the initial condition

$$x|_{t=t_0} = x_0, \quad y|_{t=t_0} = y_0, \tag{2.2}$$

and we write $\mathfrak{T}_\varepsilon$ for the phase trajectory in the (x,y) plane corresponding to the solution (2.1). The next section of this chapter will be devoted to the asymptotic approximation of $\mathfrak{T}_\varepsilon$ for $\varepsilon \to 0$.

We shall construct the phase trajectory $\mathfrak{T}_0$ of the degenerate system (1.2) starting at Q_0 (Fig. 29). Under the assumptions of Sec. 1, a heuristic analysis (cf. Chapter I, Sec. 4) of the motion of the phase point of (1.1) in the (x,y) plane is easily carried out. This analysis shows that *in any bounded part of the phase plane, for sufficiently small* ε, *the trajectory* $\mathfrak{T}_\varepsilon$ *is close to the trajectory* $\mathfrak{T}_0$. This is proved by indicating a neighborhood of $\mathfrak{T}_0$ of small but finite (i.e., not depending on ε) width, such that the phase-velocity vectors of (1.1) are directed towards the interior of the neighborhood at all points of the boundary for sufficiently small ε. The construction of this neighborhood, which clearly uses specific properties of the phase-velocity vector of the nondegenerate system is clear from the example shown in Fig. 30.

We also have a stronger assertion: *In any finite part of the phase plane, the curve* $\mathfrak{T}_0$ *can serve as a zeroth asymptotic approximation for the trajectory* $\mathfrak{T}_\varepsilon$, *i.e.,* $\mathfrak{T}_\varepsilon \to \mathfrak{T}_0$ *when* $\varepsilon \to 0$ *uniformly on any segment of* $\mathfrak{T}_0$ *of finite length.*

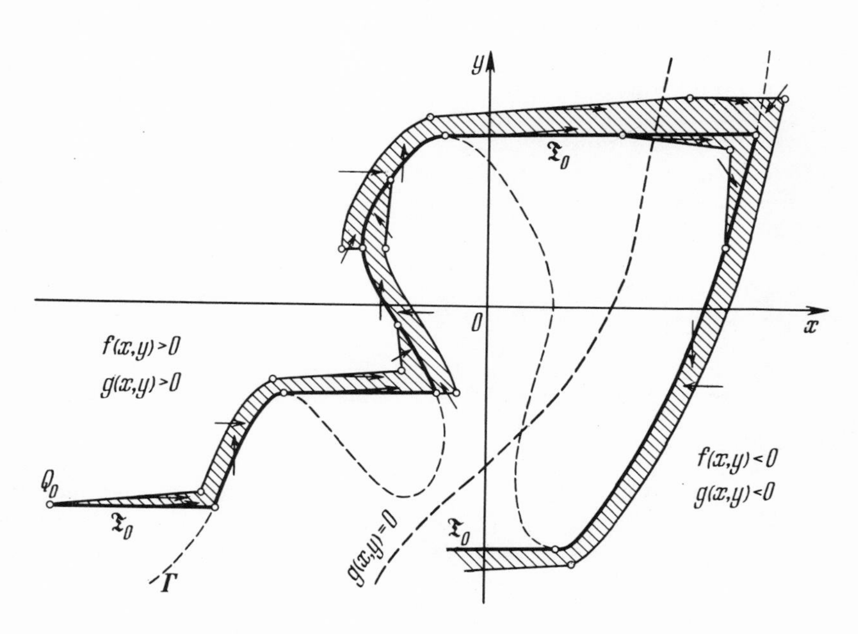

Fig. 30

This can be seen by using simple inequalities and carrying out a more precise construction of $\mathfrak{T}_0$, so that it can be shown that the neighborhood contracts on $\mathfrak{T}_0$ when $\varepsilon \to 0$ [19, 3]. We do not dwell on this here, because this result will be an automatic conclusion of our determination of asymptotic approximations of the trajectory $\mathfrak{T}_\varepsilon$.

An analysis of the structure of the phase-velocity vector of the nondegenerate system leads to the conclusion (Chapter 1, Sec. 4) that the phase point of (1.1) traverses parts of $\mathfrak{T}_\varepsilon$, at a great distance from Γ, considerably more rapidly than it traverses parts of $\mathfrak{T}_\varepsilon$, close to Γ. We therefore divide $\mathfrak{T}_\varepsilon$ into the following four types of parts (Fig. 30):

(a) a *slow-motion part,* close to the slow-motion part of $\mathfrak{T}_0$, outside finite neighborhoods of drop points and junction points;

(b) a *junction part,* in finite neighborhoods of junction points of $\mathfrak{T}_0$;

(c) a *fast-motion part,* close to the fast-motion part of $\mathfrak{T}_0$, and outside finite neighborhoods of junction and drop points;

(d) a *drop part,* in a finite neighborhood of the drop point of $\mathfrak{T}_0$.

The construction of asymptotic approximations for a segment of $\mathfrak{T}_\varepsilon$ of arbitrary finite length can clearly be reduced to the same construction for each of the parts (it still remains to consider the *initial fast-motion part,* starting at the initial point, and the subsequent *initial drop part*). These problems are solved in the sections to follow. We shall thus obtain formulas for the calculation of a segment of $\mathfrak{T}_\varepsilon$ of arbitrary finite length to any given degree of accuracy with respect to ε. We do not investigate the parts of $\mathfrak{T}_\varepsilon$ that go to infinity.

3. Asymptotic Approximations on Slow-Motion Parts of the Trajectory

Let $\widetilde{PS}$ be a slow-motion part of $\mathfrak{T}_0$, on the stable part (1.7) of Γ (Fig. 31). Here $\tilde{P}(\tilde{p}_1, \tilde{p}_2)$ is a drop point of $\mathfrak{T}_0$, and $S(s_1, s_2)$ is a junction point on the boundary of the stable part (1.7). We write $S^*(s_1^*, s_2^*)$ for the second boundary point of this part ($s_2^* = \pm\infty$ is not excluded). Clearly

$$s_2^* \operatorname{sign} g(S) < \tilde{p}_2 \operatorname{sign} g(S) < s_2 \operatorname{sign} g(S).$$

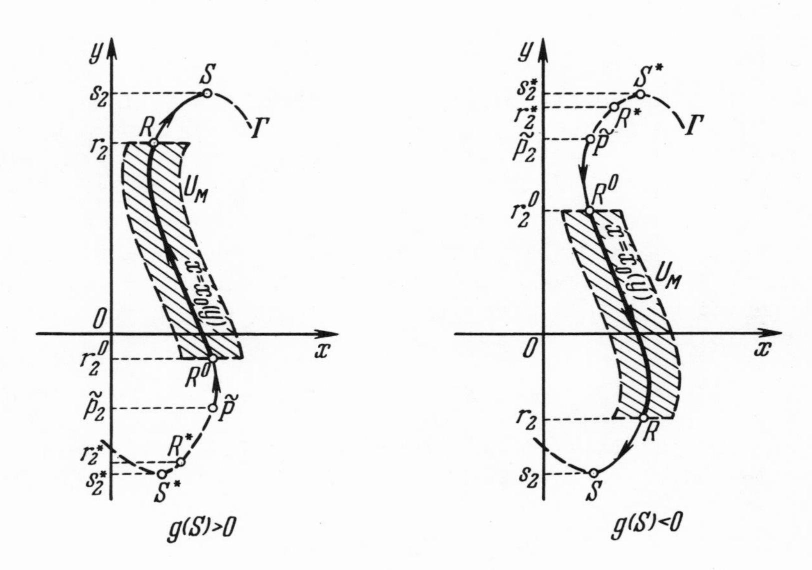

Fig. 31

On $\tilde{P}S$, take a point $R(r_1,r_2)$ at a small finite distance from the junction point S, and a point $R^0(r_1^0,r_2^0)$ at a small finite distance from the drop point $\tilde{P}$. The equation of the part R^0R of $\mathfrak{T}_0$ can be written as

$$x=x_0(y), \quad r_2^0 \operatorname{sign} g(S) \leqslant y \operatorname{sign} g(S) \leqslant r_2 \operatorname{sign} g(S). \tag{3.1}$$

Also let $R^*(r^*,r^*)$ be a point of S^*S at a small finite distance from the nonregular point S^*. By virtue of the stability of the part (1.7) on R^*R [and *a fortiori* on the curve (3.1)]

$$f'_x(x_0(y),\, y) \leqslant -k < 0. \tag{3.2}$$

Since there is no stable equilibrium position of (1.1) on the part (1.7), there is a finite neighborhood U_M of the curve (3.1) determined by the inequalities

$$r_2^0 \operatorname{sign} g(S) \leqslant y \operatorname{sign} g(S) \leqslant r_2 \operatorname{sign} g(S), \quad |x-x_0(y)| \leqslant \delta$$

(Fig. 31), lying completely in the attraction region of the stable part (1.7), at each point of which*

---

*Various constants whose values are of no importance in the reasoning will often be denoted by the same letter.

$$|g(x, y)| \geqslant k > 0. \tag{3.3}$$

Since, if ε is small enough, $\mathfrak{T}_\varepsilon$ is close to $\mathfrak{T}_0$ (Sec. 2 and Fig. 30), the trajectory $\mathfrak{T}_\varepsilon$ enters U_M through a part $y = r_2^0$ of the boundary and leaves through a part $y = r_2$ of the boundary. The segment of $\mathfrak{T}_\varepsilon$, in U_M will be called the *slow-motion part*.

By virtue of (3.3), the coordinate y on the slow-motion part of $\mathfrak{T}_\varepsilon$ varies monotonically with increasing t. Hence the equation of this part can be written, not in the parametric form (2.1), but as

$$x = x_{\mathfrak{T}}(y, \varepsilon), \quad r_2^0 \operatorname{sign} g(S) \leqslant y \operatorname{sign} g(S) \leqslant r_2 \operatorname{sign} g(S), \tag{3.4}$$

where y is the independent variable. It is clear that the function in (3.4) is a solution of the equation

$$\varepsilon \frac{dx}{dy} = \frac{f(x, y)}{g(x, y)} \equiv h(x, y), \tag{3.5}$$

which has a smooth right side and which is equivalent to the system (1.1). We use (3.5) to find an asymptotic expansion for the slow-motion part (3.4).

We construct the formal power series

$$x = x_0(y) + \varepsilon x_1(y) + \ldots + \varepsilon^n x_n(y) + \ldots \tag{3.6}$$

so that it *formally* satisfies (3.5) for y between s_2^* and s_2. The coefficients in the series are determined as follows. We substitute the series (3.6) in both sides of (3.5), formally differentiating the series term by term to obtain the derivative on the left and expanding the right side in powers of ε. This expansion is conveniently obtained by using the formula

$$F\left(z_0 + \sum_{n=1}^{\infty} \varepsilon^n z_n\right) = F(z_0) + \sum_{n=1}^{\infty} \varepsilon^n \sum_{k=1}^{n} \frac{F^{(k)}(z_0)}{k!} \sum_{\substack{i_1 + \ldots + i_k = n, \\ i_j \geqslant 1}} z_{i_1} \ldots z_{i_k}, \tag{3.7}$$

which is easily verified. We then compare coefficients of powers of ε to obtain the relations

$$h(x_0(y), y) = 0,$$

$$x_0'(y) = x_1(y) h_x'(x_0(y), y), \tag{3.8}$$

$$x'_{n-1}(y)=x_n(y)h'_x(x_0(y),\ y)+$$
$$+\Phi_{n-1}(x_0(y),\ x_1(y),\dots,\ x_{n-1}(y)),\quad n\geqslant 2.$$

The first relation here is satisfied identically if (1.7) is used for $x_0(y)$ [see (1.8) and (3.5)]. Since

$$h'_x(x_0(y),\ y)=\frac{f'_x(x_0(y),\ y)}{g(x_0(y),\ y)}, \tag{3.9}$$

and thus $h'_x(x_0(y),y) \neq 0$ by virtue of (3.2), the remaining relations (3.8), which are linear algebraic equations, successively and uniquely determine the coefficients $x_i(y)$, i = 1, 2, ...:

$$x_1(y)=-\frac{f'_y(x_0(y),\ y)\,g(x_0(y),\ y)}{f'^2_x(x_0(y),\ y)} \tag{3.10}$$

[here we have used (1.9)]:

$$x_n(y)=\frac{1}{f'_x(x_0(y),\ y)}\Bigg\{x'_{n-1}(y)\,g(x_0(y),\ y)-$$
$$-\sum_{\nu=2}^{n}\frac{f_x^{(\nu)}(x_0(y),\ y)}{\nu!}\sum_{\substack{i_1+\dots+i_\nu=n,\\ i_j\geqslant 1}}x_{i_1}(y)\dots x_{i_\nu}(y)+$$
$$+\sum_{\substack{i+k=n-1,\\ i\geqslant 0,\ k\geqslant 1}}x'_i(y)\sum_{\nu=1}^{k}\frac{g_x^{(\nu)}(x_0(y),\ y)}{\nu!}\sum_{\substack{i_1+\dots+i_\nu=k,\\ i_j\geqslant 1}}x_{i_1}(y)\dots x_{i_\nu}(y)\Bigg\},\qquad n\geqslant 2. \tag{3.11}$$

Our formulas show that the functions $x_i(y)$, i = 0, 1, 2,..., are defined and are smooth functions of y on the interval between s_2^* and s_2. It should be stressed that each of these functions can be effectively calculated in terms of only values of the right sides in (1.1) and some of their derivatives on the part S^*S of Γ. All of these functions can, of course, be considered on any interval of values of y in the interval between s_2^* and s_2, for example, on the interval between r_2^0 and r_2.

In Sec. 4 we shall see that a *partial sum*

$$X_n(y,\ \varepsilon)=x_0(y)+\varepsilon x_1(y)+\dots+\varepsilon^n x_n(y) \tag{3.12}$$

of the series in (3.6) is, for $\varepsilon \to 0$, *an asymptotic ap-*

proximation for the slow-motion part (3.4) of $\mathfrak{T}_\varepsilon$:

$$x_{\mathfrak{T}}(y,\varepsilon)=X_{N-1}(y,\varepsilon)+O(\varepsilon^N),$$
$$r_2^0 \operatorname{sign} g(S) \leqslant y \operatorname{sign} g(S) \leqslant r_2 \operatorname{sign} g(S), \tag{3.13}$$

here N is any positive integer. It follows that

$$x_{\mathfrak{T}}(y,\varepsilon)=x_0(y)+O(\varepsilon),$$
$$r_2^0 \operatorname{sign} g(S) \leqslant y \operatorname{sign} g(S) \leqslant r_2 \operatorname{sign} g(S),$$

i.e., the part (3.1) of $\mathfrak{T}_0$ can serve as a zeroth approximation for the slow-motion part of $\mathfrak{T}_\varepsilon$ under consideration. We also have the higher-order approximation [cf. (3.10)]

$$x_{\mathfrak{T}}(y,\varepsilon)=x_0(y)+\varepsilon\frac{-f'_y(x_0(y),y)\,g(x_0(y),y)}{f'^2_x(x_0(y),y)}+O(\varepsilon^2),$$
$$r_2^0 \operatorname{sign} g(S) \leqslant y \operatorname{sign} g(S) \leqslant r_2 \operatorname{sign} g(S). \tag{3.14}$$

4. Proof of the Asymptotic Representations of the Slow-Motion Part

We now establish a relation between the functions $X_n(y,\varepsilon)$ obtained in Sec. 3 and the trajectories of (1.1), i.e., we verify that there is an asymptotic representation of the type (3.13) for intervals of trajectories of this system. We then prove (3.13) for the slow-motion part (3.4) of $\mathfrak{T}_\varepsilon$.

Lemma 1. If the points R^* and R on S^*S are fixed, then for $n = 0, 1, 2, \ldots$, there is a constant $M_n > 0$ such that

$$|x_n(y)| \leqslant M_n,$$
$$r_2^* \operatorname{sign} g(S) \leqslant y \operatorname{sign} g(S) \leqslant r_2 \operatorname{sign} g(S). \tag{4.1}$$

The proof is by induction. We use (3.10), (3.11), inequality (3.2), and the fact that $f(x,y)$ and $g(x,y)$ and their derivatives are bounded near the part (1.7) outside finite neighborhoods of the nonregular points S^* and S.

Lemma 2. If ε is small enough, there is a constant $C_n > 0$ such that the derivative, in accordance with (1.1), is positive at each point of the curve

$$\mathcal{K}_1(y,x)\equiv x-X_n(y,\varepsilon)+C_n\varepsilon^{n+1}=0, \tag{4.2}$$
$$r_2^* \operatorname{sign} g(S) \leqslant y \operatorname{sign} g(S) \leqslant r_2 \operatorname{sign} g(S),$$

and negative at each point of the curve

$$\mathcal{K}_2(y, x) \equiv x - X_n(y, \varepsilon) - C_n\varepsilon^{n+1} = 0,$$
$$r_2^* \operatorname{sign} g(S) \leqslant y \operatorname{sign} g(S) \leqslant r_2 \operatorname{sign} g(S). \quad (4.3)$$

It follows from (1.1) that the derivative

$$\left.\frac{d\mathcal{K}_1}{dt}\right|_{(1.1)} = \left.\frac{dx}{dt}\right|_{(1.1)} - \frac{dX_n(y, \varepsilon)}{dy} \cdot \left.\frac{dy}{dt}\right|_{(1.1)} =$$

$$= \frac{1}{\varepsilon} g(X_n(y, \varepsilon) - C_n\varepsilon^{n+1}, y)\left\{h(X_n(y, \varepsilon) - C_n\varepsilon^{n+1}, y) - \varepsilon \frac{dX_n(y, \varepsilon)}{dy}\right\}$$

at each point of the curve (4.2). We now find the dominant term of this expression for sufficiently small ε, assuming that

$$r_2^* \operatorname{sign} g(S) \leqslant y \operatorname{sign} g(S) \leqslant r_2 \operatorname{sign} g(S).$$

It follows from (4.1) that

$$g(X_n(y, \varepsilon) - C_n\varepsilon^{n+1}, y) = g(x_0(y), y) + O(\varepsilon),$$
$$h(X_n(y, \varepsilon) - C_n\varepsilon^{n+1}, y) = h(X_n(y, \varepsilon), y) - C_n\varepsilon^{n+1}h'_x(x_0(y), y) + O(\varepsilon^{n+2}), \quad (4.4)$$

where the remainder term converges uniformly for $\varepsilon \to 0$ for y between r_2^* and r_2. Moreover, the same process of constructing the functions $x_i(y)$ [see (3.8)] and (3.7)] implies that

$$\varepsilon \frac{dX_n(y, \varepsilon)}{dy} = h(X_n(y, \varepsilon), y) + \varepsilon^{n+1}x_{n+1}(y)h'_x(x_0(y), y) + O(\varepsilon^{n+2}). \quad (4.5)$$

It follows from (4.4), (4.5), and (3.9) that

$$\left.\frac{d\mathcal{K}_1}{dt}\right|_{(1.1)} = -\varepsilon^n\{C_n + x_{n+1}(y)\} f'_x(x_0(y), y) + O(\varepsilon^{n+1}).$$

The first term is dominant for sufficiently small ε and, if we choose the constant $C_n > 0$ to be large enough, (4.1) and (3.2) imply that this term is strictly positive for y between r_2^* and r_2. Hence it can be ensured that

$$\left.\frac{d\mathcal{K}_1}{dt}\right|_{(1.1)} \geqslant k > 0$$

at all points of the curve (4.2).

It can be proved similarly that, if ε is small enough, we can ensure that

$$\left.\frac{d\mathcal{K}_2}{dt}\right|_{(1.1)} \leqslant -k < 0$$

at all points of the curve (4.3) by taking $C_n > 0$ to be sufficiently large.

Let $R^0(r_1^0, r_2^0)$ be any interior point of the arc R^*R. We shall prove the following assertion concerning the possibility of obtaining an asymptotic approximation for intervals of trajectories of (1.1) by using partial sums of the series in (3.6).

Theorem 1. Suppose that (1.7) is a stable part of the curve Γ, the functions $X_n(y, \varepsilon)$, $n \geq 0$, are given by (3.12), and $x = x(y, \varepsilon)$ is the solution of (3.5) with an initial value for $y = r_2^0$ satisfying the condition

$$|x(r_2^0, \varepsilon) - X_n(r_2^0, \varepsilon)| < C\varepsilon^{n+1}, \quad C > 0, \tag{4.6}$$

for sufficiently small ε. Then, if ε is small enough, this solution is defined on the interval determined by the inequalities

$$r_2^0 \operatorname{sign} g(S) \leqslant y \operatorname{sign} g(S) \leqslant r_2 \operatorname{sign} g(S) \tag{4.7}$$

and has the representation

$$\begin{gathered} x(y, \varepsilon) = X_n(y, \varepsilon) + \mathfrak{R}_n(y, \varepsilon), \\ r_2^0 \operatorname{sign} g(S) \leqslant y \operatorname{sign} g(S) \leqslant r_2 \operatorname{sign} g(S), \end{gathered} \tag{4.8}$$

and

$$|\mathfrak{R}_n(y, \varepsilon)| < C_n \varepsilon^{n+1}, \qquad C_n = \text{const} > 0, \tag{4.9}$$

uniformly on the interval (4.7).

This theorem follows directly from Lemma 2. In fact, consider the strip π between the curves (4.2) and (4.3), and let the constant $C_n > 0$ be so large that the assertion of Lemma 3 holds (Fig. 32). If the solution $x = x(y, \varepsilon)$ of (3.5) satisfies the initial condition (4.6), we assume (increasing the constant C_n if necessary) that the initial point $(x(r_2^0, \varepsilon), r_2^0)$ corresponding to a trajectory of (1.1) is in the strip π. By virtue of Lemma 3, no trajectory of (1.1) starting from any interior point of π intersects the curve $\mathcal{K}_1(y, x) = 0$ or the curve $\mathcal{K}_2(y, x) = 0$. Moreover, there is no equilibrium position of (1.1) in π. Hence $x = x(y, \varepsilon)$ is defined on the whole interval (4.7), and its graph is in the interior of π. We conclude that we have the representation (4.8) and that inequality (4.9) holds.

Theorem 1 proves that we have the asymptotic representation (3.13) for the segments of trajectories of (1.1) under consideration.

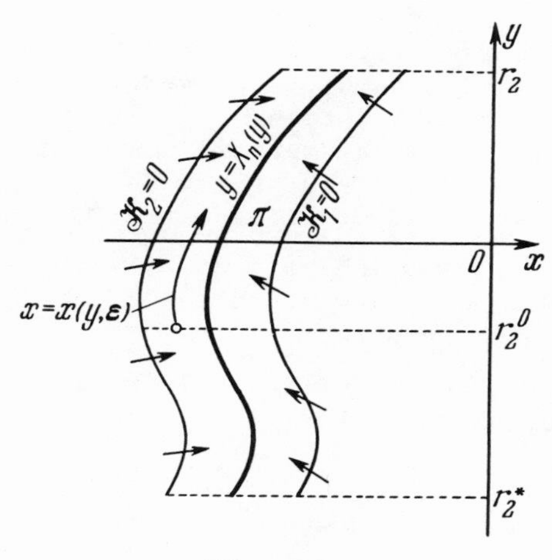

Fig. 32

Now take $R^0(r_1^0, r_2^0)$ to be an arc $\tilde{P}S$ at a small finite distance from the drop point $\tilde{P}$, and let A be the point of intersection of the trajectory $\mathfrak{T}_\varepsilon$, defined in Sec. 2 with the straight line $y = r_2^0$, lying on the boundary of the region U_M (Fig. 33; see also Fig. 31). This point, which is an initial point for a slow-motion part, is simultaneously the end of the preceding drop part. The drop part of $\mathfrak{T}_\varepsilon$

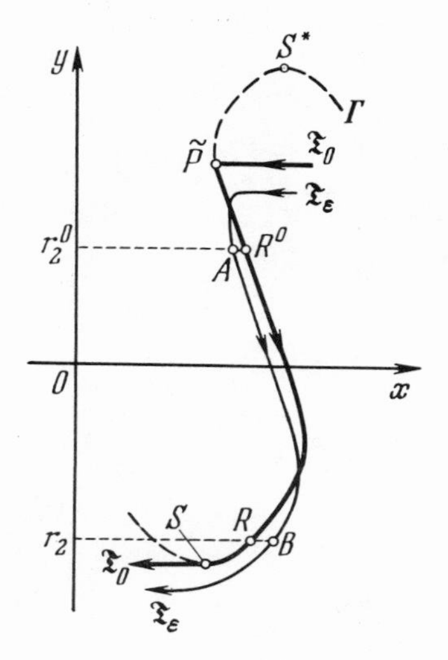

Fig. 33

will be examined in detail in Secs. 19-22, where we shall see that the abscissa of the point A satisfies (4.6). In other words, condition (4.6), for the function (3.4), is actually satisfied for all integers $n \geq 0$; hence Theorem 1 can be applied to obtain asymptotic approximations for the slow-motion part of $\mathfrak{T}_\varepsilon$. This proves (3.13).

It is clear that (3.13) can be used to calculate approximately the slow-motion part of $\mathfrak{T}_\varepsilon$ with any given accuracy (with error of order at most ε^α, $\alpha > 0$) for $\varepsilon \to 0$, uniformly on the whole part.

Theorem 2. If ε is sufficiently small, we have the following representation for the slow-motion part (3.4) of $\mathfrak{T}_\varepsilon$:

$$x_{\mathfrak{T}}(y, \varepsilon) = X_{]a]}(y, \varepsilon) + O(\varepsilon^a),$$

$$r_2^0 \operatorname{sign} g(S) \leqslant y \operatorname{sign} g(S) \leqslant r_2 \operatorname{sign} g(S);$$

here $a > 0$ is arbitrary, and we write $]a]$ for the largest integer strictly smaller than a.* No integration of the nondegenerate system (1.1) is necessary in the calculation of the function $X_{]a]}(y, \varepsilon)$.

5. Local Coordinates in the Neighborhood of a Junction Point

Relations (3.10) and (3.11) show that the functions $x_i(y)$, $i = 1, 2, \ldots$, increase in absolute value without limit when $y \to s_2$. Hence the series in (3.6) cannot be considered to be an asymptotic expansion for $\mathfrak{T}_\varepsilon$ on the whole range of the variable y between r_2^0 and s_2. Construction of asymptotic approximations of the part of $\mathfrak{T}_\varepsilon$, close to the junction point $S(s_1, s_2)$ requires the use of a new type of series.

To derive asymptotic representations of this part, it is convenient first to replace, in a small finite neighborhood of the junction point S, the variables x and y by local coordinates ξ and η, such that the junction point becomes the origin and the adjacent part of Γ is transformed into an arc of a quadratic parabola. This transformation is carried out as follows.

In the neighborhood of the junction point $S(s_1, s_2)$, the equation of Γ can be written in the form (1.13). Consider

*The expression $[a]$ usually denotes the largest integer not exceeding a.

the function

$$y=\varphi(x)\equiv x\sqrt{\left|\frac{f''_x(S)}{2f'_y(S)}\right|-\frac{xY'''(s_1+\vartheta x)}{6}\operatorname{sign}[f''_x(S)f'_y(S)]};\qquad(5.1)$$

it is clear that $y = \varphi(x)$ is a smooth real-valued function in the neighborhood of $x = 0$, and that

$$\varphi(0)=0,\qquad \varphi'(0)=\sqrt{\left|\frac{f''_x(S)}{2f'_y(S)}\right|}\qquad(5.2)$$

Since $\varphi'(0) > 0$ [see (1.12)], $\varphi'(x) > 0$ everywhere on some interval $\tilde{\rho}_{-1}\leqslant x\leqslant\tilde{\rho}_{+1}$, where $\tilde{\rho}_{-1}<0$, and $\tilde{\rho}_{+1}>0$. Hence the function $y=\varphi(x)$, $\tilde{\rho}_{-1}\leqslant x\leqslant\tilde{\rho}_{+1}$, has a smooth inverse $x = \psi(y)$, $-q_*\leqslant y\leqslant q^*$. We assume that $x = \psi(y)$ is defined on a small interval for $|y|\leqslant q$, where q is a small positive number *(not larger than unity)*, independent of ε; the corresponding interval $\rho_{-1}\leqslant x\leqslant\rho_{+1}$ of definition of $y = \varphi(x)$ can be found directly.

Consider a small finite neighborhood U_S of S:

$$s_1+\rho_{-1}\leqslant x\leqslant s_1+\rho_{+1},\quad s_2-q\leqslant y\leqslant s_2+q.$$

We assume that, in U_S, the functions $f''_x(x,y)$, $f'_x(x,y)$, and $g(x,y)$ have constant sign (this can always be arranged by decreasing q), and we replace x and y by new coordinates ξ and η defined as follows:

$$\begin{cases}\xi=\varphi(x-s_1)\operatorname{sign}f''_x(S),\\ \eta=(y-s_2)\operatorname{sign}g(S);\end{cases}\qquad\begin{cases}x=s_1+\psi(\xi\operatorname{sign}f''_x(S)),\\ y=s_2+\eta\operatorname{sign}g(S).\end{cases}\qquad(5.3)$$

Equations (5.3) establish a smooth one-to-one relation between the region U_S of the (x,y) plane and a small finite neighborhood U_0 of the origin in the (ξ,η) plane defined by the relations

$$|\xi|\leqslant q,\quad |\eta|\leqslant q$$

(Fig. 34). Clearly, the point S corresponds to the origin $\xi = 0$, $\eta = 0$, and it is easily verified, by using (1.15), that the part of the curve (1.13) corresponding to the interval $s_1+\rho_{-1}\leqslant x\leqslant s_1+\rho_{+1}$ of variation of x is converted, in the new coordinates, into an arc of a parabola:

$$\eta=-\xi^2,\quad -q\leqslant\xi\leqslant q;\qquad(5.4)$$

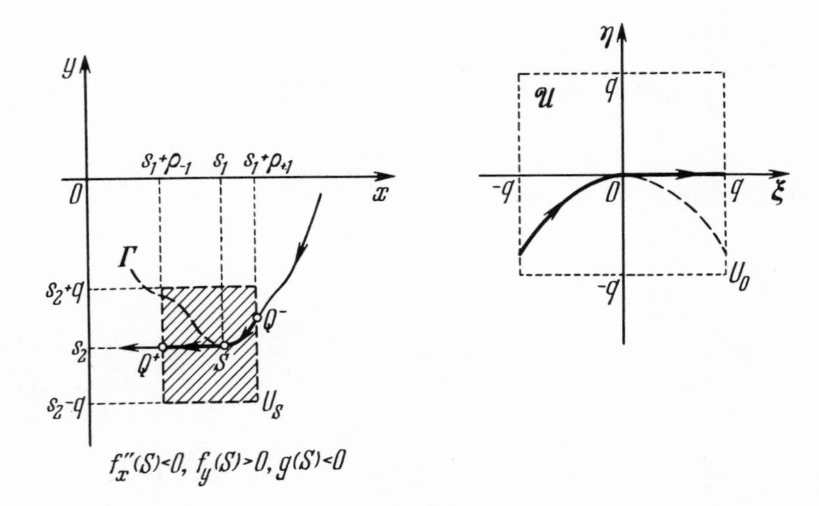

Fig. 34

the identity (1.1) becomes

$$f(s_1+\psi(\xi \operatorname{sign} f''_x(S)),\ s_2-\xi^2 \operatorname{sign} g(S)) \equiv 0, \quad |\xi| \leqslant q. \tag{5.5}$$

Investigation of each of the possible cases shown in Figs. 26 and 27 shows that the part, adjacent to S, of the stable part of Γ becomes, in the new coordinates, the parabolic arc (5.4) corresponding to the interval $-q \leqslant \xi < 0$ of variation of ξ.

In the new coordinates, the nondegenerate system (1.1) becomes

$$\begin{cases} \varepsilon\dot{\xi} = \dfrac{\xi^2+\eta}{\alpha(\xi, \eta)}, \\ \dot{\eta} = \beta(\xi, \eta), \end{cases} \tag{5.6}$$

where

$$\alpha(\xi, \eta) \equiv \frac{(\xi^2+\eta)\,\psi'(\xi \operatorname{sign} f''_x(S)) \operatorname{sign} f''_x(S)}{f(s_1+\psi(\xi \operatorname{sign} f''_x(S)),\ s_2+\eta \operatorname{sign} g(S))},$$
$$\beta(\xi, \eta) \equiv |g(s_1+\psi(\xi \operatorname{sign} f''_x(S),\ s_2+\eta \operatorname{sign} g(S))|. \tag{5.7}$$

The function $\alpha(\xi,\eta)$ is defined everywhere in U_0, including the curve (5.4). In fact, by virtue of (5.5),

$$f(s_1+\psi(\xi \operatorname{sign} f''_x(S)),\ s_2+\eta \operatorname{sign} g(S)) =$$
$$= (\xi^2+\eta) f'_y(s_1+\psi(\xi \operatorname{sign} f''_x(S)),\ s_2-\xi^2 \operatorname{sign} g(S) +$$
$$+ \vartheta(\xi^2+\eta) \operatorname{sign} g(S)) \operatorname{sign} g(S)$$

(here the constant $\vartheta \in [0,1]$), and $f'_y(x,y)$ does not vanish in U_S. Plainly $\alpha(\xi,\eta)$ and $\beta(\xi,\eta)$ are smooth functions of their arguments. Finally, it is easily seen [by using (1.15) and the fact that $g(x,y) \neq 0$ in U_S] that

$$\alpha(\xi,\eta) \geqslant k > 0, \quad \beta(\xi,\eta) \geqslant k > 0 \tag{5.8}$$

at each point of U_0; hence, in particular, $\alpha(0,0) > 0$ and $\beta(0,0) > 0$.

To simplify the notation, we always denote the values of $\alpha(\xi,\eta)$, $\beta(\xi,\eta)$, etc., and their derivatives $\alpha'_\eta(\xi,\eta)$, $\alpha''_{\xi\eta}(\xi,\eta)$, $\beta''_{\xi\eta}(\xi,\eta)$, etc., for $\xi = 0$, $\eta = 0$ by α, β, α'_η, $\alpha''_{\xi\eta}$, $\beta''_{\xi\eta}$, etc.

Let $Q^-(q_1^-,q_2^-)$ be the point with the abscissa

$$q_1^- = s_1 + \rho_{-\operatorname{sign} f''_x(S)} = s_1 + \psi(-q \operatorname{sign} f''_x(S)),$$

and let $Q^+(q_1^+,q_2^+)$ be the point with the abscissa

$$q_1^+ = s_1 + \rho_{\operatorname{sign} f''_x(S)} = s_1 + \psi(q \operatorname{sign} f''_x(S))$$

(Q^- is in the stable part of Γ adjacent to the junction point S, and Q^+ is on the horizontal segment through S; see Fig. 34.) It can be verified directly that the equation of the part Q^-Q^+ of $\mathfrak{T}_0$, independently of which representation in Figs. 26 and 27 suits the case under consideration, can be written as follows in the coordinates ξ,η:

$$\eta = \eta_0(\xi) = \begin{cases} -\xi^2, & -q \leqslant \xi < 0, \\ 0, & 0 \leqslant \xi \leqslant q. \end{cases} \tag{5.9}$$

Since, for sufficiently small ε, the trajectory $\mathfrak{T}_\varepsilon$ is close to $\mathfrak{T}_0$ (see Sec. 2 and Fig. 30), the trajectory $\mathfrak{T}_\varepsilon$ enters U_S through the part

$$x = q_1^-, \quad q_2^- \operatorname{sign} g(S) \leqslant y \operatorname{sign} g(S) \leqslant s_2 \operatorname{sign} g(S)$$

of the boundary, and leaves through the part

$$x = q_1^+, \quad s_2 \operatorname{sign} g(S) \leqslant y \operatorname{sign} g(S) \leqslant s_2 \operatorname{sign} g(S)\, q$$

of the boundary, and *is everywhere above the part* Q^-Q^+ *of* $\mathfrak{T}_0$ *if* $g(S) > 0$, *and is everywhere below this part if* $g(S) < 0$. The section of $\mathfrak{T}_\varepsilon$ in U_S will be called the *junction part*. It follows from (5.3) that this part corresponds to the segment of the trajectory

$$\xi = \xi(t,\varepsilon), \quad \eta = \eta(t,\varepsilon) \tag{5.10}$$

of the system (5.6) on the part $\mathcal{U}$ of the region U_0 above the curve (5.9) (see Fig. 34).

6. Asymptotic Approximations of the Trajectory on the Initial Part of a Junction

Since $\xi^2 + \eta > 0$ and $\alpha(\xi,\eta) > 0$ in the region $\mathcal{U}$ [see (5.8)], on the junction part of the trajectory $\mathfrak{T}_\varepsilon$ the coordinate ξ is a monotonically increasing function of the time t. Hence the equation of this part need not be written in the parametric form (5.10), but can be written as

$$\eta = \eta_{\mathfrak{T}}(\xi, \varepsilon), \qquad -q \leqslant \xi \leqslant q, \tag{6.1}$$

with the independent variable ξ. We see that the function (6.1) is a solution of the equation

$$\frac{d\eta}{d\xi} = \varepsilon \frac{\gamma(\xi, \eta)}{\xi^2 + \eta}, \tag{6.2}$$

where

$$\gamma(\xi, \eta) \equiv \alpha(\xi, \eta)\beta(\xi, \eta) \tag{6.3}$$

is a smooth function defined everywhere in U_0; equation (6.2) is equivalent to the system (5.6). It follows from (5.8) that, at any point of U_0 (and thus in $\mathcal{U}$),

$$\gamma(\xi, \eta) \geqslant k > 0, \tag{6.4}$$

so that $\gamma = \gamma(0,0) > 0$.

It is not possible to obtain an asymptotic expansion for $\varepsilon \to 0$ of the junction part (6.1) as a series in positive powers of ε. We thus begin by using (6.2) to seek an asymptotic expansion of $\mathfrak{T}_\varepsilon$ at the initial junction part, i.e., for $-q \leq \xi \leq \sigma_1$, where $\sigma_1 > 0$ is small when ε is small, that is, $\sigma_1 \to 0$ when $\varepsilon \to 0$. In other words, we are interested in asymptotic expansions of the function

$$\eta = \eta_{\mathfrak{T}}(\xi, \varepsilon), \qquad -q \leqslant \xi \leqslant -\sigma_1, \tag{6.5}$$

coinciding with the function (6.1) on the interval on which they are both defined.

We construct the formal power series

$$\eta = \eta_0(\xi) + \varepsilon\eta_1(\xi) + \ldots + \varepsilon^n\eta_n(\xi) + \ldots \tag{6.6}$$

so that it *formally* satisfies (6.2) for $-q \leq \xi < 0$. The coefficients in this series are determined as follows: For $\eta_0(\xi)$ we use the function (5.9) on the interval $-q \leq \xi < 0$. This is a natural choice, because (5.9) is the equation of a part of $\mathfrak{T}_0$ in the coordinates ξ, η. Furthermore, we use (6.6) in (6.2), with formal differentiation term by term on the left side and a series expansion in powers of ε on the right side by means of the formula

$$\frac{1}{z_0+\sum_{n=1}^{\infty}\varepsilon^n z_n}=\frac{1}{z_0}+\sum_{n=1}^{\infty}\varepsilon^n\sum_{k=1}^{n}\frac{(-1)^k}{z_0^{k+1}}\sum_{\substack{i_1+\ldots+i_k=n\\ i_j\geqslant 1}} z_{i_1}\ldots z_{i_k}, \qquad (6.7)$$

which is a special case of (3.7). We now equate coefficients of powers of ε to obtain the following relations:

$$\begin{aligned} &\eta_0'(\xi)=\frac{\gamma(\xi,\eta_0(\xi))}{\eta_1(\xi)},\\ &\eta_1'(\xi)=-\frac{\gamma(\xi,\eta_0(\xi))}{\eta_1^2(\xi)}\eta_2(\xi)+\gamma_\eta'(\xi,\eta_0(\xi)), \qquad (6.8)\\ &\eta_{n-1}'(\xi)=-\frac{\gamma(\xi,\eta_0(\xi))}{\eta_1^2(\xi)}\eta_n(\xi)+\Phi_{n-1}(\eta_0(\xi),\eta_1(\xi),\ldots,\eta_{n-1}(\xi)), \quad n\geqslant 2. \end{aligned}$$

By virtue of the choice of $\eta_0(\xi)$, the first equation here means that we can write

$$\eta_1(\xi)=-\frac{\gamma(\xi,-\xi^2)}{2\xi}. \qquad (6.9)$$

Since

$$-\frac{\gamma(\xi,\eta_0(\xi))}{\eta_1^2(\xi)}=-\frac{4\xi^2}{\gamma(\xi,-\xi^2)}\neq 0,$$

the other relations in (6.8), which are linear algebraic equations, have a unique solution for the coefficients $\eta_i(\xi)$, $i = 2, 3, \ldots$:

$$\begin{aligned} &\eta_2(\xi)=-\frac{\gamma^2(\xi,-\xi^2)-\xi\gamma(\xi,-\xi^2)\gamma_\xi'(\xi,-\xi^2)}{8\xi^4},\\ &\eta_n(\xi)=\frac{1}{2\xi}\Bigg\{\sum_{\nu=1}^{n-1}\eta_{n-\nu}(\xi)\,\eta_\nu'(\xi)- \qquad (6.10)\\ &-\sum_{\nu=1}^{n-1}\frac{\gamma_\eta^{(\nu)}(\xi,-\xi^2)}{\nu!}\sum_{\substack{i_1+\ldots+i_\nu=n-1,\\ i_j\geqslant 1}}\eta_{i_1}(\xi)\ldots\eta_{i_\nu}(\xi)\Bigg\}, \quad n\geqslant 2. \end{aligned}$$

These formulas show that the functions $\eta_i(\xi)$, $i = 0, 1, 2, \ldots$, are defined as smooth functions for $-q \le \xi < 0$. We stress that each of these functions can be effectively calculated in terms of values of $\gamma(\xi, \eta)$ and some of their derivatives on the interval $-q \le \xi < 0$ of the curve (5.9).

In Secs. 7 and 16 we shall see that a *partial sum* of the series (6.6),

$$H_n(\xi, \varepsilon) = \eta_0(\xi) + \varepsilon\eta_1(\xi) + \ldots + \varepsilon^n\eta_n(\xi), \tag{6.11}$$

is an asymptotic representation, for $\varepsilon \to 0$, of the section (6.5) of the trajectory $\mathfrak{T}_\varepsilon$:

$$\eta_{\mathfrak{T}}(\xi, \varepsilon) = H_{N_1-1}(\xi, \varepsilon) + O(\varepsilon^{N_1 - \lambda_1(3N_1-2)}), \quad -q \leqslant \xi \leqslant -\sigma_1; \tag{6.12}$$

where N_1 is any positive integer and

$$\sigma_1 = \varepsilon^{\lambda_1}, \quad 0 < \lambda_1 < 1/3. \tag{6.13}$$

7. The Relation between Asymptotic Representations and Actual Trajectories in the Initial Junction Section

We shall demonstrate the validity of asymptotic expansions of the type (6.12) and (6.13) for the separate sections or trajectories of the system (5.6).

Lemma 3. For $n = 0, 1, 2, \ldots$, there is a constant $M_n > 0$ such that

$$|\xi^{3n-2}\eta_n(\xi)| \leqslant M_n, \quad -q \leqslant \xi < 0. \tag{7.1}$$

By using (6.9), (6.10), and the fact that $\gamma(\xi,\eta)$ and its derivatives are bounded in $\mathcal{U}$, we can prove by induction that

$$|\xi^{3n-2+m}\eta_n^{(m)}(\xi)| \leqslant M_{n,m}, \quad -q \leqslant \xi < 0,$$

where m is any nonnegative integer. Inequality (7.1) is the special case of this inequality for $m = 0$.

More detailed calculations show that, for all positive integers, the function $\eta_n(\xi)$ has the asymptotic expansion

$$\eta_n(\xi) = \frac{1}{\xi^{3n-2}} \sum_{k=0}^{\infty} \eta_k^n \xi^k, \quad \xi \to -0, \tag{7.2}$$

and we can obtain a recurrence formula permitting the successive calculation of the coefficients η_k^n. Since

$$\eta_0^1 = -\frac{\gamma}{2}, \quad \eta_0^n = -\frac{3n-4}{4}\sum_{\nu=1}^{n-1} \eta_0^\nu \eta_0^{n-\nu}, \quad n \geqslant 2, \tag{7.3}$$

we have $\eta_0^n < 0$ for all positive integers n [see (6.4)]; hence

$$\begin{aligned} \eta_n(\xi) &\to +\infty \quad \text{when} \quad \xi \to -0, \text{ if } n \text{ is odd}, \\ \eta_n(\xi) &\to -\infty \quad \text{when} \quad \xi \to -0, \text{ if } n \text{ is even}. \end{aligned} \tag{7.4}$$

Lemma 4. For sufficiently small ε, there is a constant $C_n > 0$ such that the derivative calculated in accordance with (6.2), at each point of the curve

$$\mathcal{K}_1(\xi, \eta) \equiv \eta - H_n(\xi, \varepsilon) + C_n \frac{\varepsilon^{n+1}}{\xi^{3n+1}} = 0, \quad -q \leqslant \xi \leqslant -\sigma_1, \tag{7.5}$$

is positive for odd n and negative for even n, and

$$\mathcal{K}_2(\xi, \eta) \equiv \eta - H_n(\xi, \varepsilon) - C_n \frac{\varepsilon^{n+1}}{\xi^{3n+1}} = 0, \qquad -q \leqslant \xi \leqslant -\sigma_1, \tag{7.6}$$

is negative for odd n and positive for even n.

It follows from (6.2) that the derivative is given at points of the curve (7.5) by the formula

$$\frac{d\mathcal{K}_1}{d\xi}\bigg|_{(6.2)} = \frac{d\eta}{d\xi}\bigg|_{(6.2)} - \frac{dH_n(\xi, \varepsilon)}{d\xi} + \frac{d}{d\xi}\frac{C_n\varepsilon^{n+1}}{\xi^{3n+1}} =$$

$$= \varepsilon \frac{\gamma\left(\xi, H_n(\xi, \varepsilon) - C_n \frac{\varepsilon^{n+1}}{\xi^{3n+1}}\right)}{\xi^2 + H_n(\xi, \varepsilon) - C_n \frac{\varepsilon^{n+1}}{\xi^{3n+1}}} - \frac{dH_n(\xi, \varepsilon)}{d\xi} - \frac{(3n+1)\, C_n\varepsilon^{n+1}}{\xi^{3n+2}}.$$

We shall find the dominant term of this expression for small ε, assuming that $-q \leq \xi \leq -\sigma_1$, where $\sigma_1 = \varepsilon^{\lambda_1}$, $0 < \lambda_1 < 1/3$ [see (6.13)]. By virtue of (6.9) and (7.1)

$$\varepsilon \frac{\gamma\left(\xi, H_n(\xi, \varepsilon) - C_n \frac{\varepsilon^{n+1}}{\xi^{3n+1}}\right)}{\xi^2 + H_n(\xi, \varepsilon) - C_n \frac{\varepsilon^{n+1}}{\xi^{3n+1}}} = \varepsilon \frac{\gamma(\xi, H_n(\xi, \varepsilon))}{\xi^2 + H_n(\xi, \varepsilon)} +$$

$$+ \varepsilon^n \frac{4C_n}{\gamma(\xi, -\xi^2)\, \xi^{3n-1}} + O(\varepsilon^{n+1-\lambda_1(3n+2)}); \tag{7.7}$$

the remainder term in this formula is uniformly bounded for $-q \leq \xi \leq -\sigma_1$. The same procedure applied to $\eta_i(\xi)$ [see (6.8)], and the use of (3.7) and (6.7), yield

$$\frac{dH_n(\xi,\varepsilon)}{d\xi} = \varepsilon\frac{\gamma(\xi, H_n(\xi,\varepsilon))}{\xi^2 + H_n(\xi,\varepsilon)} - \varepsilon^n\frac{4\xi^2}{\gamma(\xi,-\xi^2)}\eta_{n+1}(\xi) + O(\varepsilon^{n+1-\lambda_1(3n+2)}); \quad (7.8)$$

it follows from (7.1) and (6.13) that, for $-q \leq \xi \leq -\sigma_1$, the remainder term here is strictly of higher order than the second term for $\varepsilon \to 0$.

It follows from (7.7) and (7.8) that

$$\left.\frac{d\mathcal{K}_1}{d\xi}\right|_{(6.2)} = \varepsilon^n\frac{4}{\gamma(\xi,-\xi^2)\,\xi^{3n-1}}\{\xi^{3n+1}\eta_{n+1}(\xi) + C_n\} + O(\varepsilon^{n+1-\lambda_1(3n+2)}).$$

If $0 < \lambda_1 < 1/3$, then for sufficiently small ε the first term on the right is dominant and, by virtue of (7.1), when the constant $C_n > 0$ is large enough this term is strictly positive for odd n and strictly negative for even n for $-q \leq \xi \leq -\sigma_1 < 0$. Hence, at all points of the curve (7.5),

$$\left.\frac{d\mathcal{K}_1}{d\xi}\right|_{(6.2)} \geqslant k > 0, \quad \text{if } n \text{ is odd,}$$

and

$$\left.\frac{d\mathcal{K}_1}{d\xi}\right|_{(6.2)} \leqslant -k < 0, \quad \text{if } n \text{ is even.}$$

It can be proved similarly that, for sufficiently small ε,

$$\left.\frac{d\mathcal{K}_2}{d\xi}\right|_{(6.2)} \leqslant -k < 0, \quad \text{if } n \text{ is odd,}$$

and

$$\left.\frac{d\mathcal{K}_2}{d\xi}\right|_{(6.2)} \geqslant k > 0, \quad \text{if } n \text{ is even}$$

at the points of the curve (7.6); this is proved by taking $C_n > 0$ to be sufficiently large.

We now prove the following result concerning the possibility of obtaining an asymptotic approximation for segments of trajectories of the system (5.6) by using partial sums of the series in (6.6).

Theorem 3. Let λ_1 be any number satisfying $0 < \lambda_1 < 1/3$, let $H_n(\xi,\varepsilon)$, $n \geq 0$, be defined by (6.11), and let $\eta = \eta(\xi,\varepsilon)$ be the solution of Eq. (6.2) with an initial value for $\xi = -q$, $q = \text{const} > 0$, satisfying

$$|\eta(-q,\varepsilon) - H_n(-q,\varepsilon)| < C\varepsilon^{n+1}, \quad C > 0, \quad (7.9)$$

for sufficiently small ε. Then, for sufficiently small ε, this solution exists for $-q \leq \xi \leq -\sigma_1$, $\sigma_1 = \varepsilon^{\lambda_1}$, and has the representation

$$\eta(\xi, \varepsilon) = \mathrm{H}_n(\xi, \varepsilon) + \mathfrak{R}_n(\xi, \varepsilon), \quad -q \leqslant \xi \leqslant -\sigma_1, \tag{7.10}$$

and

$$|\mathfrak{R}_n(\xi, \varepsilon)| < C_n \varepsilon^{n+1-\lambda_1(3n+1)}, \quad C_n = \mathrm{const} > 0 \tag{7.11}$$

uniformly on the interval $-q \leq \xi \leq -\sigma_1$.

Theorem 3 follows directly from Lemma 4. It is clear that, if n is odd, the curve (7.6) is always above the curve (7.5), while if n is even the curve (7.5) is always above the curve (7.6). Consider the strip π between these curves, and assume that the constant $C_n > 0$ is so large that the assertion of Lemma 4 holds. If the solution $\eta = \eta(\xi, \varepsilon)$ of (6.2) satisfies the initial condition (7.9), then (possibly by increasing C_n) we assume that the initial point $(-q, \eta(-q, \varepsilon))$ is in π. Since Lemma 4 implies that no solution of (6.2) starting at any point of π intersects either the curve $\mathcal{K}_1(\xi, \eta) = 0$ or the curve $\mathcal{K}_2(\xi, \eta) = 0$, the function $\eta = \eta(\xi, \varepsilon)$ is defined for $-q \leq \xi \leq -\sigma_1$, and its graph is in the interior of π. This proves (7.10) and (7.11).

It is important that there is a direct relation between coefficients of the asymptotic series (3.6) and (6.6). If the series in (6.6) is rewritten by using (5.3) in the coordinates x, y, the resulting series

$$\begin{aligned} y = s_2 + \eta_0(\varphi(x - s_1)\operatorname{sign} f''_x(S))\operatorname{sign} g(S) + \\ + \varepsilon\eta_1(\varphi(x - s_1)\operatorname{sign} f''_x(S))\operatorname{sign} g(S) + \ldots \\ \ldots + \varepsilon^n\eta_n(\varphi(x - s_1)\operatorname{sign} f''_x(S))\operatorname{sign} g(S) + \ldots \end{aligned} \tag{7.12}$$

is formally the transformation of the series (3.6), and the nth coefficient in (7.12) can be expressed in terms of the coefficients in (3.6) with numbers not exceeding n. To prove this, we must establish the following relations between coefficients in the series (3.6) and in the series (6.6):

$$\begin{gathered} x_0(s_2 + \eta_0(\xi)\operatorname{sign} g(S)) \equiv s_1 + \psi(\xi \operatorname{sign} f''_x(S)), \\ \eta_1(\xi)\, x'_0(s_2 + \eta_0(\xi)\operatorname{sign} g(S))\operatorname{sign} g(S) + x_1(s_2 + \eta_0(\xi)\operatorname{sign} g(S)) \equiv 0, \\ \ldots\ldots\ldots\ldots\ldots\ldots\ldots\ldots ; \end{gathered}$$

this can be done by using (1.11) and (5.5) and the expressions for the coefficients in the series under consideration [see (3.10), (3.11), (6.9), and (6.10)].

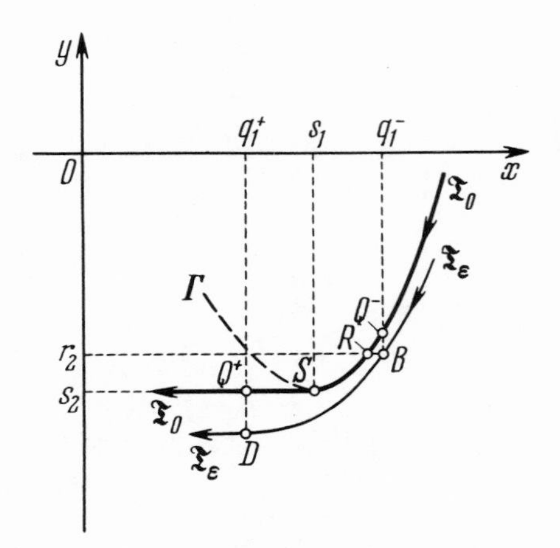

Fig. 35

We now prove that the function (6.5) describing the trajectory $\mathfrak{T}_\varepsilon$ in the initial junction section satisfies (7.9) for all integral $n \geq 0$. We choose q in accordance with the conditions in Sec. 5, and thus determine a point $Q^-(q_1^-, q_2^-)$ on the arc $\tilde{PS}$ of the trajectory $\mathfrak{T}_0$ (Fig. 35; cf. Fig. 34). Let $B(q_1^-, r_2)$ be the point of intersection of $\mathfrak{T}_\varepsilon$ with $x = q_1^-$, on the boundary of U_S, and let $R(r_1, r_2)$ be the point on $\tilde{PS}$ with the same ordinate as the point B [so that $r_1 = x_0(r_2)$]. The point B is the initial point of the junction section, and is simultaneously the end of the preceding slow-motion part (Fig. 33). It should be stressed that, although r_2 depends on ε, the point R, for sufficiently small ε, is at a finite distance from the junction point S; hence results obtained in Secs. 3 and 4 can be used for the asymptotic calculation of the slow-motion part AB of the trajectory $\mathfrak{T}_\varepsilon$.

The quantity $\eta_{\mathfrak{T}}(-q, \varepsilon)$ needed for the verification of condition (7.9) is obtained as follows. We fix any integer $n \geq 0$ and, using the asymptotic representation (3.13) for the slow-motion part AB, we obtain the abscissa

$$q_1^- = x_{\mathfrak{T}}(r_2, \varepsilon) = X_n(r_2, \varepsilon) + O(\varepsilon^{n+1})$$

of the point B. Now using (5.3) to transform to the coordinates ξ, η we obtain the relation

$$s_1 + \psi(-q \operatorname{sign} f_x''(S)) = X_n(s_2 + \eta_{\mathfrak{T}}(-q, \varepsilon) \operatorname{sign} g(S), \varepsilon) + O(\varepsilon^{n+1}), \tag{7.13}$$

which determines the quantity $\eta_{\mathfrak{T}}(-q, \varepsilon)$ with the required accuracy. But the relations between the coefficients in the series (3.6) and (6.6) obtained previously show that the transformation of (7.13) yields

$$\eta_{\mathfrak{T}}(-q,\ \varepsilon)=\mathrm{H}_n(-q,\ \varepsilon)+O(\varepsilon^{n+1}).$$

Thus condition (7.9) for the function (6.5) is satisfied for integers $n \geq 0$.

8. Special Variables for the Junction Section

Relations (7.4) show that the series in (6.6) in positive powers of ε cannot be considered as an asymptotic expansion for $\mathfrak{T}_\varepsilon$ on the whole interval $-q \leq \xi \leq 0$ of variation of ξ. For example $\eta_1(\xi)$ increases unboundedly when $\xi \to -0$ [see (6.8)]; hence the difference $\eta_{\mathfrak{T}}(\xi,\varepsilon) - \eta_0(\xi)$ characterizing the deviation of $\mathfrak{T}_\varepsilon$ from $\mathfrak{T}_0$, is not of order ε for $\varepsilon \to 0$ uniformly on $-q \leq \xi \leq 0$.

However, the order of this difference can be found; it is $\varepsilon^{2/3}$ for $\varepsilon \to 0$ uniformly on some interval of values of ξ including $\xi = 0$. To prove this and to obtain an asymptotic representation of $\mathfrak{T}_\varepsilon$ on the interval of values of ξ referred to, we make a new variable change.

We transform from the variables ξ, η, t to variables u, v, τ by means of the formulas

$$\xi=\mu u, \qquad \eta=\mu^2 v, \qquad t=\mu^2\tau, \tag{8.1}$$

where we have used the notation

$$\mu^3=\gamma\varepsilon. \tag{8.2}$$

It is clear that μ can be considered as a new *parameter* that is positive [see (6.4)] and is small when ε is small, but is not of higher order of smallness than ε (it is, in fact, of order $\varepsilon^{1/3}$) for $\varepsilon \to 0$. The new time τ is a fast time; the third relation in (8.1) shows that a finite interval of values of t corresponds to an interval of variation of τ with length of order $1/\mu^2$. The first two equations in (8.1) define a one-to-one relation between the small finite neighborhood U_0 of the origin in the (ξ,η) plane and a region U_0^* of the (u,v) plane that becomes infinitely large when $\varepsilon \to 0$. The origin of the (ξ,η) plane is transformed into the origin of the (u,v) plane, and the curve (5.9) is transformed into the curve

$$v=\begin{cases} -u^2, & -q/\mu\leqslant u<0, \\ 0, & 0\leqslant u\leqslant q/\mu. \end{cases} \tag{8.3}$$

In the new variables, the system (5.6) becomes

$$\begin{cases} \dfrac{du}{d\tau} = \gamma \dfrac{u^2+v}{\alpha(\mu u, \mu^2 v)}, \\ \dfrac{dv}{d\tau} = \beta(\mu u, \mu^2 v); \end{cases} \tag{8.4}$$

in this new system no derivative is multiplied by a small parameter. The right-hand sides of (8.4) are defined and are smooth functions everywhere in U_0^*. By virtue of (8.1), the junction part (5.10) of the trajectory $\mathfrak{T}_\varepsilon$ corresponds to the segment of the trajectory

$$u = u(\tau, \mu), \qquad v = v(\tau, \mu) \tag{8.5}$$

of (8.4) in the part $\mathfrak{U}^*$ of the region $\overset{*}{U_0}$ above the curve (8.3). Equation (8.3) describes the part (5.9) of the trajectory $\mathfrak{T}_0$, in the coordinates u,v; however, the curve (8.3) does not satisfy the system obtained from (8.4) by putting $\mu = 0$.

9. A Riccati Equation

The equation

$$\frac{dv}{du} = \frac{1}{u^2+v} \tag{9.1}$$

arises in the construction of the zeroth approximation of trajectories of the system (8.4). If u is assumed to be a function of the independent variable v, (9.1) can be written as

$$\frac{du}{dv} = v + u^2; \tag{9.2}$$

this is a *special Riccati equation*, whose solution can be expressed in terms of special functions [21]. We use the solution to find the general behavior of integral curves of Eq. (9.1).

Replace u in (9.2) by the new variable $e(v)$ defined by the relation

$$u(v) = -\frac{1}{e(v)} \cdot \frac{de(v)}{dv}. \tag{9.3}$$

The function $e(v)$ satisfies the linear differential equation

$$\frac{d^2e}{dv^2} + ve = 0, \tag{9.4}$$

which is called *Airy's equation.* The change of variables

$$w=\frac{e}{\sqrt{v}},\qquad z=\frac{2}{3}v^{3/2},$$

converts this equation into the *Bessel equation*

$$z^2\frac{d^2w}{dz^2}+z\frac{dw}{dz}+\left[z^2-\left(\frac{1}{3}\right)^2\right]w=0,$$

which has a fundamental solution system consisting of Bessel functions of the first kind of order $-1/3$ and $1/3$, $J_{-1/3}(z)$ and $J_{1/3}(z)$. Hence the general solution of (9.4) can be written as

$$e(v)=C_1\sqrt{v}\,J_{-1/3}\left(\frac{2}{3}v^{3/2}\right)+C_2\sqrt{v}\,J_{1/3}\left(\frac{2}{3}v^{3/2}\right),\tag{9.5}$$

where C_1 and C_2 are arbitrary constants.

Returning to Riccati's equation (9.2), we use (9.5) and (9.3) to express its general solution in terms of Bessel functions,

$$u=-\sqrt{v}\,\frac{J_{-2/3}\left(\frac{2}{3}v^{3/2}\right)-cJ_{2/3}\left(\frac{2}{3}v^{3/2}\right)}{cJ_{-1/3}\left(\frac{2}{3}v^{3/2}\right)+J_{1/3}\left(\frac{2}{3}v^{3/2}\right)};\tag{9.6}$$

the constant c can take any real value (including ∞). Formula (9.6) is useful for $v \geq 0$ because, for nonnegative values of the argument, it directly gives real values of $u = u(v,c)$. For negative values of the argument, i.e., when $-v > 0$, another representation of the general solution is preferable,

$$u=-\sqrt{-v}\,\frac{I_{-2/3}\left(\frac{2}{3}(-v)^{3/2}\right)-cI_{2/3}\left(\frac{2}{3}(-v)^{3/2}\right)}{cI_{-1/3}\left(\frac{2}{3}(-v)^{3/2}\right)-I_{1/3}\left(\frac{2}{3}(-v)^{3/2}\right)};\tag{9.7}$$

here $I_\nu(z)$ is the modified Bessel function of the first kind of order ν; it gives real values of $u = u(v,c)$ for $v < 0$.

We now examine the graphs of the functions of the family given by (9.6) or, equivalently, by (9.7). For all c (including $c = \infty$), a function $u = u(v,c)$ of this family is defined on the whole v axis, except at points where the denominator in the right-hand side of (9.6) or (9.7) vanishes. For $0 \leq c \leq 1$, one such point is on the half-line $v \leq 0$ and there is a countable sequence of these points, tending to infinity, on the half-line $v > 0$; for other values of c there is only a countable sequence of these points, tending to infinity, on the positive half-line.* If we approach such a point from the left, $u(v,c)$ tends to $+\infty$, and if we approach it from the right the function tends to $-\infty$. Every function of the family $u = u(v,c)$ except $u = u(v,1)$ has the asymptotic representation

$$u(v, c) = \sqrt{-v}\,[1 + O((-v)^{-3/2})], \quad v \to -\infty, \quad c \neq 1,$$

while**

$$u(v, 1) = -\sqrt{-v}\,[1 + O((-v)^{-3/2})], \quad v \to -\infty.$$

In other words, the graphs of all functions $u = u(v,c)$, $c \neq 1$, for $v \to -\infty$ approach the upper branch of the parabola $\Pi(v,u) \equiv v + u^2 = 0$, which is the isocline of zero for Eq. (9.2), and $u = u(v,1)$ is the *only* function whose value approximates the lower branch of this parabola when $v \to -\infty$. We also note that the curves $u = u(v,c)$ have points of inflection at their intersections with the curve $1 + 2uv + 2u^3 = 0$, which is formed of two branches Γ_1 and Γ_2.

Figure 36 shows integral curves of Riccati's equation (9.2). The previous analysis implies that (9.2) has only solutions of the following three types:

(1) a continuum of curves, each of which is defined on its specific interval, half-infinite to the left, and approaches the curve $u = (-v)^{1/2}$ when $v \to -\infty$, and tends to $+\infty$ when the variable approaches the right end of its interval of definition;

(2) a continuum of curves, each of which is defined on its specific finite interval, tends to $-\infty$ when the variable approaches the left end of its interval of definition, and tends to $+\infty$ when the variable approaches the right end;

(3) a *single* curve defined on an interval, half-infinite to the left, approaching the curve $u = -(-v)^{1/2}$ when $v \to -\infty$ and tending to $+\infty$ when the variable approaches the right ends of its interval of definition.

---

*This is because the function $cI_{-1/3}(z) - I_{1/3}(z)$ has a single zero for $z \geq 0$ if $0 \leq c \leq 1$ and has no zeros in this interval for other c, while $cJ_{-1/3}(z) + J_{1/3}(z)$ has a countable number of zeros for $c > 0$ (all these zeros are simple) [66].

**This is proved by using asymptotic formulas for $I_\nu(z)$ and $I_{-\nu}(z) - I_\nu(z)$ for $z \to \infty$.

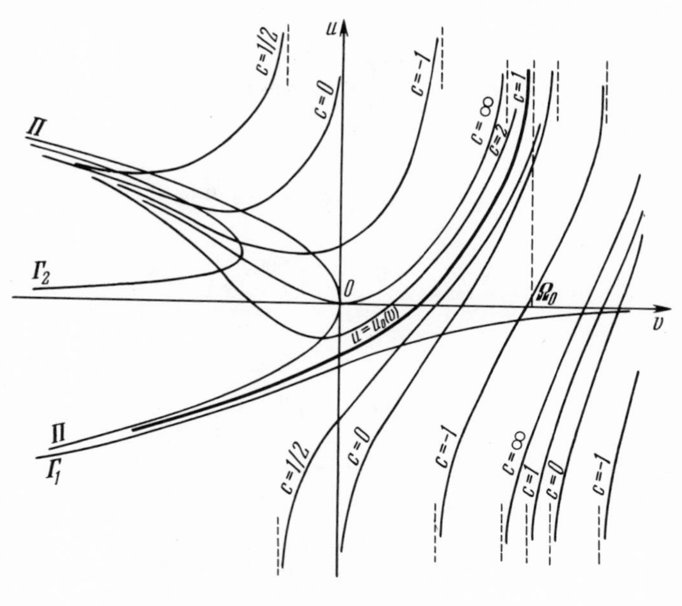

Fig. 36

Figure 36 shows that this last curve divides the region filled with curves of the first type from the region filled with curves of the second type. Such a solution of Riccati's equation is called a *dividing* solution.

The dividing solution $u = u_0(v)$ of (9.2) is defined for $-\infty < v < \Omega_0$, where Ω_0 is the smallest positive zero of the denominator in (9.6) for $c = 1$. This solution increases monotonically from $-\infty$ to $+\infty$ and is convex. For $-\infty < v \leq 0$, it is between the lower branch of the parabola Π and the curve Γ_1.

The mirror reflection of the curves in Fig. 36 in the bisector of the first and third quadrants consists of integral curves of (9.1), and in this mirror reflection the dividing curve $u = u_0(v)$ of Riccati's equation (9.2) is the solution

$$v = v_0(u) \tag{9.8}$$

of (9.1). This function (Fig. 37) will play an important part in the construction of asymptotic representations of trajectories of the system (8.4). Although the function (9.8) cannot be obtained explicitly in terms of known functions, its

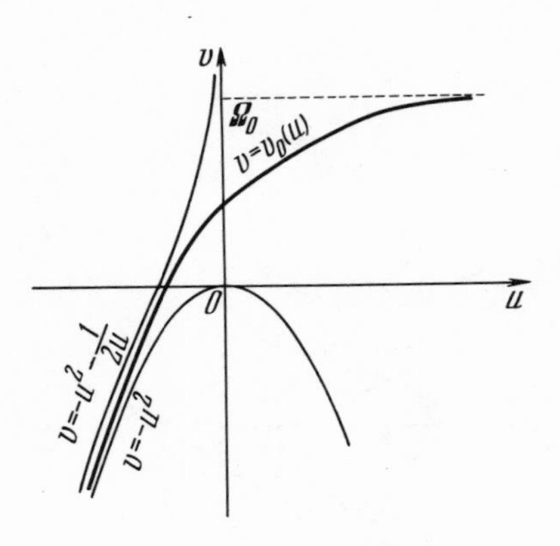

Fig. 37

properties can be investigated by using the expression for its inverse $u = u_0(v)$ and Eq. (9.1) itself.

It is clear that the function (9.8) is defined on the whole line $-\infty < u < \infty$, is monotonically increasing and convex above, and

$$\lim_{u \to +\infty} v_0(u) = \Omega_0 > 0. \tag{9.9}$$

It should be noted that (9.8) is the *only* solution of (9.1) approaching the left branch of the parabola $v = -u^2$ when $u \to -\infty$. Moreover,

$$-u^2 < v_0(u) < -u^2 - \frac{1}{2u}, \quad -\infty < u < 0. \tag{9.10}$$

We shall need further asymptotic formulas for the function (9.8) for large negative and large positive values of the argument. These formulas can be directly obtained from (9.1):

$$v_0(u) = -u^2 - \frac{1}{2u} - \frac{1}{8u^4} + O\left(\frac{1}{u^7}\right), \quad u \to -\infty; \tag{9.11}$$

$$v_0(u) = \Omega_0 - \frac{1}{u} + O\left(\frac{1}{u^3}\right), \qquad u \to +\infty. \tag{9.12}$$

10. Asymptotic Approximations for the Trajectory in the Neighborhood of a Junction Point

Since, in $\mathcal{U}^*$, $u^2 + v > 0$ and $\alpha(\mu u, \mu^2 v) > 0$, $\gamma > 0$ [see (5.8) and (6.4)], we conclude that the coordinate u increases monotonically for increasing time τ on the section determined by (8.5). Hence the equation of this section can be written, not in the parametric form (8.5), but as

$$v = v_{\mathfrak{T}}(u, \mu), \qquad -q/\mu \leqslant u \leqslant q/\mu, \tag{10.1}$$

with the independent variable u; the function in (10.1) is obtained from (6.1) by the change of variables (8.1), (8.2); it is clear that this function is a solution of the equation

$$\frac{dv}{du} = \frac{\tilde{\gamma}(\mu u, \mu^2 v)}{u^2 + v}, \tag{10.2}$$

where

$$\tilde{\gamma}(\xi, \eta) = \frac{\gamma(\xi, \eta)}{\gamma(0, 0)} = \frac{1}{\gamma}\gamma(\xi, \eta) \tag{10.3}$$

is a smooth function everywhere in U_0. Equation (10.2) is equivalent to the system (8.4). It follows from (10.3) that $\gamma = \tilde{\gamma}(0,0) = 1$.

It turns out that we cannot obtain an asymptotic expansion for $\mu \to 0$ (i.e., for $\varepsilon \to 0$) of the function (10.1) as a series in positive integral powers of μ. We therefore use Eq. (10.2) and seek an asymptotic expansion of (10.1), not on the whole interval of its definition, but only on an interval $-\omega_1 \leq u \leq \omega_2$, where $\omega_1 > 0$ and $\omega_2 > 0$ are quantities that are arbitrarily large for arbitrarily small μ, i.e., $\omega_i \to +\infty$, $i = 1, 2$, when $\varepsilon \to 0$.

Consider the formal power series

$$v = v_0(u) + \mu v_1(u) + \ldots + \mu^n v_n(u) + \ldots, \tag{10.4}$$

formally satisfying (10.2) for $-\infty < u < \infty$. We use this series in (10.2), and formally expand the right side of the equation in powers of μ. The expansion is easily obtained from (3.7) and (6.6),

$$\frac{\tilde{\gamma}\left(\mu u, \mu^2 \sum_{k=0}^{\infty} \mu^k v_k\right)}{u^2 + v_0 + \sum_{k=1}^{\infty} \mu^k v_k} = \frac{1}{u^2 + v_0} + \sum_{k=1}^{\infty} \mathfrak{M}_k \mu^k; \tag{10.5}$$

here the coefficient of μ^k, $k \geq 1$, is*

$$\mathfrak{M}_k = \frac{\mathfrak{L}_k}{u^2+v_0} + \sum_{l=1}^{k} \mathfrak{L}_{k-l} \sum_{\nu=1}^{l} \frac{(-1)^\nu}{(u^2+v_0)^{\nu+1}} \sum_{\substack{i_1+\ldots+i_\nu=l,\\ i_j \geqslant 1}} v_{i_1} \ldots v_{i_\nu},$$

$$\mathfrak{L}_m = \frac{\tilde{\gamma}_\xi^{(m)}}{m!} u^m + \sum_{n=2}^{m} \frac{u^{m-n}}{(m-n)!} \sum_{\nu=1}^{n} \frac{\tilde{\gamma}_{\xi\eta}^{(m-n)\,(\nu)}}{\nu!} \sum_{\substack{i_1+\ldots+i_\nu=n,\\ i_j \geqslant 2}} v_{i_1-2} \ldots v_{i_\nu-2}. \tag{10.6}$$

We now equate coefficients of powers of μ on the left and right to obtain

$$\frac{dv_0(u)}{du} = \frac{1}{u^2+v_0(u)},$$
$$\frac{dv_1}{du} = \mathfrak{M}_1, \ \ldots, \ \frac{dv_n}{du} = \mathfrak{M}_n, \ \ldots \tag{10.7}$$

Relations (10.6) show that the coefficient $\mathfrak{M}_k$ in the expansion (10.5) can be expressed in terms of the functions $v_0(u)$, $v_1(u)$, ..., $v_{k-1}(u)$, $v_k(u)$; the function $v_k(u)$ occurs only *once* in the expression for this coefficient — in the term corresponding to $l = k$ and $\nu = 1$. Hence, if we use the notation

$$\Phi_k(u) = \mathfrak{M}_k + \frac{v_k(u)}{[u^2+v_0(u)]^2}, \quad k = 1, 2, \ldots, \tag{10.8}$$

the relations (10.7), from the second relation on, can be written as

$$\frac{dv_1(u)}{du} + \frac{v_1(u)}{[u^2+v_0(u)]^2} = \Phi_1(u) \equiv \frac{\tilde{\gamma}'_\xi u}{u^2+v_0(u)},$$

$$\frac{dv_n(u)}{du} + \frac{v_n(u)}{[u^2+v_0(u)]^2} = \Phi_n(u), \quad n \geqslant 1. \tag{10.9}$$

The first equality in (10.7) implies that $v_0(u)$ is a solution of (9.1), and we note that Eq. (9.1) was obtained without any condition restricting us to a special solution. Hence the determination of $v_0(u)$ by using the first relation in (10.7) is not trivial.

---

*A sum in which the upper limit of summation is smaller than the lower limit will always be taken to be zero. If the condition accompanying the summation sign is not satisfied for some term, this term is taken to be zero.

For $v_0(u)$ we take the solution (9.8) of Eq. (9.1). This choice can be explained as follows. If we want the partial sums of the series in (10.4) to be asymptotic approximations of the function (10.1) for $\mu \to 0$, the coefficient of $v_0(u)$ must be considered as a zeroth approximation. It follows from results in Sec. 6 that, for $-q/\mu \leq u \leq -\sigma_1/\mu$, a zeroth approximation of the trajectory $\mathfrak{T}_\varepsilon$ is the curve $v = -u^2$. Since this function is not a solution of (9.1), we cannot use it for $v_0(u)$. However, it is natural to take $v_0(u)$ to be a solution of (9.1), which differs only slightly from the zeroth approximation indicated on $-q/\mu \leq u \leq -\sigma_1/\mu$ for small μ. Since, when $\mu \to 0$, the interval $-q/\mu \leq u \leq -\sigma_1/\mu$ tends to $-\infty$ [see (6.13) and (8.2)], it is logical to choose a solution approaching the curve $v = -u^2$ when $u \to -\infty$. As we showed in Sec. 9, Eq. (9.1) has a *unique* solution (9.8) with this property. This reasoning is, of course, only heuristic; a proof of the necessity of taking $v_0(u)$ to be the function (9.8) can be based only on a rigorous demonstration (cf. Sec. 11) that partial sums of the series (10.4) asymptotically approximate the function (10.1).

Now consider relations (10.9). Since the $\Phi_n(u)$, $n \geq 1$, can be expressed in terms of $v_0(u)$, $v_1(u), \ldots, v_{n-1}(u)$, and do not depend on the function $v_n(u)$, these relations are linearly independent differential equations directly yielding the coefficients $v_i(u)$, $i = 1, 2, \ldots$. However, there is no supplementary condition corresponding to any of these equations indicating a specific solution. We shall select the functions $v_i(u)$, $i = 1, 2, \ldots$, as follows (we omit the heuristic considerations indicating the soundness of our choice). We take $v_1(u)$ to be the following special solution of the first equation in (10.9):

$$v_1(u) = \mathscr{M}(u) \int_{-\infty}^{u} \frac{\tilde{\gamma}'_\xi \, \theta \, d\theta}{\mathscr{M}(\theta)\,[\theta^2 + v_0(\theta)]}, \tag{10.10}$$

where

$$\mathscr{M}(u) = \exp \int_{u}^{\infty} \frac{d\theta}{[\theta^2 + v_0(\theta)]^2}. \tag{10.11}$$

If the functions $v_1(u), \ldots, v_{n-1}(u)$ have already been determined, so that the $\Phi_n(u)$ [see (10.8)] are known, we take $v_n(u)$ to be the following special solution of the nth equation in (10.9):

$$v_n(u) = \mathscr{M}(u) \int_{-\infty}^{u} \frac{\Phi_n(\theta)}{\mathscr{M}(\theta)} \, d\theta, \qquad n \geq 1. \tag{10.12}$$

We shall prove below that (10.12) has a meaning for all $n \geq 1$; hence the functions $v_i(u)$, $i = 0, 1, 2, \ldots$, are defined and smooth on the whole u axis. We shall also prove that each of these functions can be effectively calculated in terms of the values of the function $\gamma(\xi,\eta)$ and some of its derivatives at the origin [the function $v_0(u)$ is in general independent of the specific form of Eq. (10.2)].

We shall prove in Secs. 11 and 16 that the partial sum

$$V_n(u, \mu) = v_0(u) + \mu v_1(u) + \ldots + \mu^n v_n(u) \tag{10.13}$$

of the series (10.4) is, for $\varepsilon \to 0$, an asymptotic approximation of the part of the solution (10.1) of Eq. (10.2) on the interval

$$-\omega_1 \leqslant u \leqslant \omega_2, \quad \omega_1 = \sigma_1/\mu, \quad \omega_2 = \sigma_2/\mu, \tag{10.14}$$

where $\sigma_1 > 0$ and $\sigma_2 > 0$ are such that, when $\varepsilon \to 0$, $\sigma_i \to 0$, and $\omega_i \to -\infty$, $i = 1, 2$ simultaneously. It is more convenient, in the statement of the result, to transform back from the variables u, v to the variables ξ, η [see (8.1)]. Then Eq. (10.2) is converted into Eq. (6.2), and the solution (10.1) on the interval (10.14) is described by the function

$$\eta = \tilde{\eta}_{\mathfrak{T}}(\xi, \mu) \equiv \mu^2 v_{\mathfrak{T}}(\xi/\mu, \mu), \quad -\sigma_1 \leqslant \xi \leqslant \sigma_2. \tag{10.15}$$

It is clear that this function coincides with the function (6.1) on their common interval of definition, and yields the equation of the part of $\mathfrak{T}_\varepsilon$ *immediately adjacent to the junction point*. We shall derive the following asymptotic representations of the part (10.15) of the trajectory $\mathfrak{T}_\varepsilon$:

$$\tilde{\eta}_{\mathfrak{T}}(\xi, \mu) = \begin{cases} -\xi^2 + O(\varepsilon^{1-\lambda_1}), \\ \mu^2 v_0\left(\dfrac{\xi}{\mu}\right) + O(\varepsilon), & -\sigma_1 \leqslant \xi \leqslant 0; \\ \mu^2 V_{N_2}\left(\dfrac{\xi}{\mu}, \mu\right) + O(\varepsilon^{1+\lambda_1 N_2}), \end{cases} \tag{10.16}$$

$$\tilde{\eta}_{\mathfrak{T}}(\xi, \mu) = \begin{cases} O(\varepsilon^{2/3}), \\ \mu^2 v_0\left(\dfrac{\xi}{\mu}\right) + O\left(\varepsilon \ln \dfrac{1}{\varepsilon}\right), & 0 \leqslant \xi \leqslant \sigma_2; \\ \mu^2 V_{N_3}\left(\dfrac{\xi}{\mu}, \mu\right) + O(\varepsilon^{1+\lambda_2 N_3}), \end{cases} \tag{10.17}$$

here N_2 and N_3 are arbitrary positive integers such that $N_2 > N_3$, and

$$\sigma_1 = \varepsilon^{\lambda_1}, \quad 0 < \lambda_1 < 1/3; \quad \sigma_2 = \varepsilon^{\lambda_2}, \quad 0 < \lambda_2 < 1/3. \tag{10.18}$$

11. The Relation between Asymptotic Approximations and Actual Trajectories in the Immediate Vicinity of a Junction Point

We now justify the asymptotic representations (10.16)-(10.18) for intervals of trajectories of the system (5.6).

Lemma 5. The functions $v_n(u)$ for $n = 1, 2, \ldots$ are defined for all real values of u and have the representations

$$v_n(u) = -\sum_{\nu=0}^{[n/2]} \frac{(-1)^\nu \tilde{\gamma}_{\xi\ \eta}^{(n-2\nu)\ (\nu)}}{\nu!\,(n-2\nu)!\,2}\, u^{n-1} + O(|u|^{n-4}) \quad \text{for } u \to -\infty; \tag{11.1}$$

$$\left.\begin{aligned} v_1(u) &= \tilde{\gamma}'_\xi \ln u + O(1) \quad \text{for } u \to +\infty,\\ v_2(u) &= \frac{\tilde{\gamma}_\xi^{(2)}}{2} u + O(1) \quad \text{for } u \to +\infty,\\ v_3(u) &= \frac{\tilde{\gamma}_\xi^{(3)}}{12} u^2 + O(\ln u) \quad \text{for } u \to +\infty,\\ v_n(u) &= \frac{\tilde{\gamma}_\xi^{(n)}}{n!\,(n-1)} u^{n-1} + O(u^{n-3}), \quad n \geqslant 4,\\ &\qquad \text{for } u \to +\infty. \end{aligned}\right\} \tag{11.2}$$

We first show that the improper integrals in (10.11) and (10.12) are convergent. The convergence of the integral in (10.11) follows directly from (9.9); hence the function $\mathcal{M}(u)$ introduced in (10.11) is defined and infinitely differentiable on the whole u axis, takes only positive values, and is monotonically decreasing. Using (9.11) and (9.12), we find that

$$\mathcal{M}(u) = \mathfrak{K} e^{-\frac{4}{3}u^3} u^2 \left[1 + O\left(\frac{1}{|u|^3}\right)\right] \quad \text{for } u \to -\infty, \tag{11.3}$$

where $\mathfrak{K}$ is a constant and*

$$\mathcal{M}(u) = 1 + O\left(\frac{1}{u^3}\right) \quad \text{for } u \to +\infty. \tag{11.4}$$

---

*It can be proved that

$$\mathfrak{K} = \exp \int_{-\infty}^{\infty} \left\{ \frac{1}{[u^2 + v_0(u)]^2} - 4u^2 \Delta_0^-(u) + \frac{2}{u} \Delta^-(u) \right\} du,$$

where

$$\Delta^-(u) = \begin{cases} 1 & \text{for } u \leqslant -1,\\ 0 & \text{for } u > -1; \end{cases} \qquad \Delta_0^-(u) = \begin{cases} 1 & \text{for } u \leqslant 0,\\ 0 & \text{for } u > 0. \end{cases}$$

Now consider the function $v_1(u)$ introduced in (10.10). The representation (9.11) implies that the right side in (10.9) has the representation

$$\Phi_1(u) = -2\tilde{\gamma}'_\xi u^2 + O\left(\frac{1}{|u|}\right) \quad \text{for} \quad u \to -\infty.$$

Since $1/\mathcal{M}(u)$ decreases exponentially when $u \to -\infty$ [see (11.3)], the integral in (10.10) obviously converges; hence $v_1(u)$ is defined and infinitely differentiable for all u. To find the asymptotic properties of $v_1(u)$ for $u \to -\infty$, we can apply l'Hopital's rule and use (11.3) to obtain (11.1) for $n = 1$. The asymptotic behavior of $v_1(u)$ for $u \to +\infty$ indicated in (11.2) can be demonstrated by integrating the asymptotic representation

$$\frac{\Phi_1(u)}{\mathcal{M}(u)} = \frac{\tilde{\gamma}'_\xi}{u} + O\left(\frac{1}{u^3}\right) \quad \text{for} \quad u \to +\infty,$$

obtained for the integral function in (10.10) by applying (11.4) and (9.12).

The lemma can now be proved by induction. Suppose that it holds for $v_1(u), \ldots, v_{n-1}(u)$; then, using (10.8) and (10.6), we obtain the following asymptotic representations for the right side of the nth equation (10.9):

$$\Phi_n(u) = -\sum_{\nu=0}^{[n/2]} \frac{2(-1)^\nu \tilde{\gamma}_\xi^{(n-2\nu)} \eta^{(\nu)}}{\nu!\,(n-2\nu)!} u^{n+1} + O(|u|^{n-2}) \quad \text{for} \quad u \to -\infty, \tag{11.5}$$

$$\Phi_n(u) = \frac{\tilde{\gamma}_\xi^{(n)}}{n!} u^{n-2} + O(u^{n-4}) \quad \text{for} \quad u \to +\infty. \tag{11.6}$$

Since $1/\mathcal{M}(u)$ decreases exponentially for $u \to -\infty$, convergence of the integral in (10.12) is obvious. It follows from (11.3) and (11.4) that (11.1) and (11.2) hold.

Lemma 6. Suppose that $F_0(u)$ is defined and smooth on the whole real line and is positive, and that $F_0(u) = 1$ for $u \leq -1$ and $F_0(u) = \ln u$ for $u \geq c$;

$$F_1(u) = \begin{cases} |x| & \text{for } |x| > \frac{1}{2}, \\ x^2 + \frac{1}{4} & \text{for } |x| < \frac{1}{2}, \end{cases} \tag{11.7}$$

$$F_n(u) = 1 + |u|^n, \quad n \geq 2.$$

Then, if ε is small enough, there is a constant $C_n > 0$ such that the derivative in (10.2) is positive at each point of the curve

$$\mathcal{K}_1(u, v) \equiv v - V_n(u, \mu) + C_n\mu^{n+1}F_n(u) = 0, \quad -\omega_1 \leqslant u \leqslant \omega_2, \tag{11.8}$$

and is negative at each point of the curve

$$\mathcal{K}_2(u, v) \equiv v - V_n(u, \mu) - C_n\mu^{n+1}F_n(u) = 0, \quad -\omega_1 \leqslant u \leqslant \omega_2. \tag{11.9}$$

At each point of the curve (11.8), the derivative in (10.2) is equal to

$$\left.\frac{d\mathcal{K}_1}{du}\right|_{(10.2)} = \left.\frac{dv}{du}\right|_{(10.2)} - \frac{dV_n(u, \mu)}{du} + \frac{d}{du} C_n\mu^{n+1}F_n(u) =$$

$$= \frac{\tilde{\gamma}(\mu u, \mu^2 V_n(u, \mu) - C_n\mu^{n+3}F_n(u))}{u^2 + V_n(u, \mu) - C_n\mu^{n+1}F_n(u)} - \frac{dV_n(u, \mu)}{du} + C_n\mu^{n+1}F_n'(u).$$

We shall find the sign of this expression for small ε, assuming that $-\omega_1 \leq u \leq \omega_2$ [see (10.14) and (10.18)].

Taylor's formula yields

$$\frac{\tilde{\gamma}(\mu u, \mu^2 V_n(u, \mu) - C_n\mu^{n+3}F_n(u))}{u^2 + V_n(u, \mu) - C_n\mu^{n+1}F_n(u)} = \frac{1}{u^2 + v_0(u)} +$$

$$+ \mathfrak{M}_1^*\mu + \ldots + \mathfrak{M}_n^*\mu^n + \mathfrak{M}_{n+1}^*\mu^{n+1} + \mathfrak{M}_{n+2}^*\mu^{n+2}. \tag{11.10}$$

We use (10.5) and (10.6), expressing v_k, $k \geq 0$, in these relations as follows:

$$v_0 = v_0(u); \qquad v_i = v_i(u), \quad i = 1, \ldots, n;$$

$$v_{n+1} = -C_nF_n(u); \quad v_j = 0, \qquad j \geqslant n+2.$$

Clearly the choice of $v_{n+1}, v_{n+2}, \ldots$ is nowhere reflected in the coefficients $\mathfrak{M}_1, \ldots, \mathfrak{M}_n$ in the expansion (10.5); hence

$$\mathfrak{M}_k^* = \mathfrak{M}_k, \quad k = 1, \ldots, n \tag{11.11}$$

(these relations naturally do not hold for $n = 0$). Furthermore, since v_{n+1} occurs in the expression for $\mathfrak{M}_{n+1}$ only once, we use (10.8) with $k = n+1$ and v_{n+1} replaced by $-C_nF_n(u)$, to obtain

$$\mathfrak{M}_{n+1}^* = \Phi_{n+1}(u) + \frac{C_nF_n(u)}{[u^2 + v_0(u)]^2}. \tag{11.12}$$

Finally, since $F_n(u)$ has the same order for $u \to -\infty$ and $u \to +\infty$ as $|v_{n+1}(u)|$ [see (1.11) and (11.2)], and $v_j = 0$, $j \geq n + 2$, the coefficient $\mathfrak{M}^*_{n+2}$ of the remainder term in (11.10) has, in the light of (10.8), the same asymptotic behavior for $u \to -\infty$ and $u \to +\infty$ as the function $\Phi_{n+2}(u)$. Thus, by virtue of (11.5) and (11.6),

$$\left|\frac{\mathfrak{M}^*_{n+2}}{u^{n+3}}\right| \leqslant M_{n+2}, \quad -\infty < u \leqslant -1, \tag{11.13}$$

$$\left|\frac{\mathfrak{M}^*_{n+2}}{u^{n}}\right| \leqslant M_{n+2}, \quad 1 \leqslant u < \infty. \tag{11.14}$$

Relations (11.10)-(11.12) and (10.8) imply that

$$\frac{d\mathcal{K}_1}{du}\bigg|_{(10.2)} = \mu^{n+1}\left\{\Phi_{n+1}(u) + \frac{C_n F_n(u)}{[u^2+v_0(u)]^2} + C_n F'_n(u) + \mu\mathfrak{M}^*_{n+2}\right\}. \tag{11.15}$$

To find the sign of this derivative, we consider separately strictly positive values of u, strictly negative values of u, and values of u close to zero. We assume that $n \geq 0$ [see (11.7)]; for $n = 1$ and $n = 0$ the reasoning is similar.

If $-\omega_1 \leq u \leq -(n/3)^{1/3}$, we use (11.13) and (10.14) to rewrite (11.15) as

$$\frac{d\mathcal{K}_1}{du}\bigg|_{(10.2)} = \mu^{n+1}|u|^{n+2}\left\{C_n\left(\frac{1}{u^2[u^2+v_0(u)]^2} - \frac{n}{|u|^3}\right) + \right.$$
$$\left. + \frac{\Phi_{n+1}(u)}{|u|^{n+2}} + \frac{C_n}{|u|^{n+2}[u^2+v_0(u)]^2} + O(\sigma_1)\right\},$$

where $O(\sigma_1)$ denotes a term of order ε^{λ_1}, $0 < \lambda_1 < 1/3$ for $\varepsilon \to 0$, uniformly on the interval under consideration. It follows from the asymptotic representation (11.5) and the relation

$$\frac{1}{u^2[u^2+v_0(u)]^2} - \frac{n}{|u|^3} > 1,$$

which holds for $u \leq -(n/3)^{1/3}$ [see (9.10)], that, if the constant $C_n > 0$ is large enough, the expression in braces is strictly positive for $-\omega_1 \leq u \leq -(n/3)^{1/3}$.

On the interval $1 \leq u \leq \omega_2$, it follows from (11.14) and (10.14) that (11.15) can be written as

$$\frac{d\mathcal{K}_1}{du}\bigg|_{(10.2)} = \mu^{n+1}u^{n-1}\left\{nC_n + \frac{\Phi_{n+1}(u)}{u^{n-1}} + \frac{C_n}{u^{n-1}[u^2+v_0(u)]^2} + \frac{C_n u}{[u^2+v_0(u)]^2} + O(\sigma_2)\right\},$$

where $O(\sigma_2)$ denotes a term of order ε^{λ_1}, $0 < \lambda_2 < 1/3$, for $\varepsilon \to 0$, uniformly on the interval under consideration. We see from the asymptotic representation (11.6) that, if $C_n > 0$ is sufficiently large, the expression in braces is strictly positive for $1 \leq u \leq \omega_2$.

Finally, it is obvious that, if $C_n > 0$ is sufficiently large, the expression in braces in (11.15) is strictly positive for $-(n/3)^{1/3} \leq u \leq 1$; hence, at each point of the curve (11.8),

$$\left.\frac{d\mathcal{K}_1}{du}\right|_{(10.2)} \geqslant k > 0.$$

The assertion of the lemma concerning the sign of the derivative can be verified similarly at each point of the curve (11.9) by using (10.2).

The following lemma determines the existence of an asymptotic approximation for segments of solutions of Eq. (10.2) in the form of partial sums of the series in (10.4).

Lemma 7. Suppose that $v = v(u,\mu)$ is a solution of (10.2) with initial value for $u = -\omega_1$ satisfying

$$|v(-\omega_1, \mu) - V_k(-\omega_1, \mu)| < C\mu^{k+1}\omega_1^k, \quad C > 0, \qquad (11.16)$$

for sufficiently small ε. Then, for sufficiently small ε and any integer $k^* \leq k$, there is a constant $C_k > 0$ such that

$$\begin{aligned} |v(u, \mu) - V_k(u, \mu)| &< C_k\mu^{k+1}F_k(u), \quad -\omega_1 \leqslant u \leqslant 0, \\ |v(u, \mu) - V_{k^*}(u, \mu)| &< C_k\mu^{k^*+1}F_{k^*}(u), \quad 0 \leqslant u \leqslant \omega_2. \end{aligned} \qquad (11.17)$$

Lemma 7 follows directly from Lemma 6. In fact, consider the strip π between the curves (11.8) and (11.9) for $n = k$ corresponding to the interval $-\omega_1 \leq u \leq 0$, and let the constant $C_k > 0$ be so large that the assertion of Lemma 6 holds. If the solution $v = v(u,\mu)$ of (10.2) satisfies the initial condition (11.16) we assume (possibly increasing the constant C_k) that the initial point $(-\omega_1, v(-\omega_1,\mu))$ is in π. By virtue of Lemma 6, no solution of (10.2) starting at a point of π intersects the curve $\mathcal{K}_1(u,v) = 0$ or the curve $\mathcal{K}_2(u,v) = 0$; hence the graph of $v = v(u,\mu)$ is in the interior of π for $-\omega_1 \leq u \leq 0$, and this proves the first of the inequalities (11.17). Now consider the strip π^* between the curves (11.8) and (11.9) corresponding to the interval $0 \leq u \leq \omega_2$ for $n = k^*$. Since $k^* \leq k$, the constant $C_k > 0$ can be taken to be so large that the point $(0,v(0,\mu))$ is in π^*, and the assertion of Lemma 6 holds. This proves the second inequality in (11.17).

Using (8.1) to transform to the coordinates ξ,η we reformulate Lemma 7 as an assertion concerning the possibility of using asymptotic approximations of the type (10.16)-(10.18) for segments of trajectories of the system (5.6).

Theorem 4. Let λ_1 and λ_2 by any numbers satisfying the inequalities $0 < \lambda_1 < 1/3$ and $0 < \lambda_2 < 1/3$, let $V_n(u,\mu)$, $n \geq 0$, be given by (10.13), and let $\eta = \tilde{\eta}(\xi,\mu)$ be a solution of (6.2) with an initial value for $\xi = -\sigma_1$, $\sigma_1 = \varepsilon^{\lambda_1}$, satisfying the condition

$$\left|\tilde{\eta}(-\sigma_1, \mu) - \mu^2 V_n\left(-\frac{\sigma_1}{\mu}, \mu\right)\right| < C\varepsilon\sigma_1^n, \; C > 0, \qquad (11.18)$$

for sufficiently small ε. Then, if ε is small enough, this solution is defined for $-\sigma_1 \leqslant \xi \leqslant \sigma_2$, $\sigma_2 = \varepsilon^{\lambda_2}$, and has the representation

$$\tilde{\eta}(\xi, \mu) = \begin{cases} \mu^2 V_n\left(\frac{\xi}{\mu}, \mu\right) + \mathfrak{R}_n(\xi, \varepsilon), & -\sigma_1 \leqslant \xi \leqslant 0, \\ \mu^2 V_{n^*}\left(\frac{\xi}{\mu}, \mu\right) + \mathfrak{R}_{n^*}(\xi, \varepsilon), & 0 \leqslant \xi \leqslant \sigma_2, \end{cases}$$

where n^* is any integer such that $0 \leq n^* \leq n$; moreover

$$|\mathfrak{R}_n(\xi, \varepsilon)| < C_n \varepsilon^{1+n\lambda_1}, \quad C_n = \text{const} > 0,$$

uniformly for $-\sigma_1 \leq \xi \leq 0$, and

$$|\mathfrak{R}_{n^*}(\xi, \varepsilon)| < \begin{cases} C_0 \varepsilon \ln \frac{1}{\varepsilon} & \text{for } n^* = 0, \\ C_{n^*} \varepsilon^{1+n^*\lambda_2} & \text{for } n^* \geqslant 1, \end{cases} \quad C_{n^*} = \text{const} > 0,$$

uniformly for $0 \leq \xi \leq \sigma_2$.

12. Asymptotic Series for the Coefficients of the Expansion Near a Junction Point

We shall need some detailed properties of the coefficients $v_n(u)$, $n = 0, 1, 2, \ldots$, in the expansion (10.4), in particular, representations of these functions by asymptotic series for $u \to -\infty$ and $u \to +\infty$. These series are obtained by applying standard methods for finding asymptotic expansions of solutions of differential equations [10, 17, 65]. Without reproducing in detail the rather long calculation, we state the final results and outline the approach used in obtaining them.

We first consider the function $v_0(u)$ introduced in Sec. 9; we note that this function is independent of the specific form of the system (8.4) or, equivalently, of the form of Eq. (10.2).

We find the function $z_0(u)$ for nonpositive values of the argument from the relation

$$v_0(u) = -u^2 + z_0(u), \qquad -\infty < u \leqslant 0. \tag{12.1}$$

The function $z_0(u)$ satisfies the equation

$$\frac{dz}{du} = 2u + \frac{1}{z}$$

obtained from (9.1), and $z_0(u)$ is the *unique* solution of this equation tending to zero when $u \to -\infty$. It follows from this equation that $z_0(u)$ has the asymptotic expansion*

$$z_0(u)^- = \sum_{k=0}^{\infty} \frac{a_k^0}{u^{3k+1}}; \tag{12.2}$$

the coefficients a_k^0 can be calculated successively from the recurrence relation

$$a_0^0 = -\frac{1}{2}, \quad a_k^0 = -\frac{3k-1}{4} \sum_{\nu=0}^{k-1} a_\nu^0 a_{k-1-\nu}^0, \quad k \geqslant 1. \tag{12.3}$$

In particular, it is easily verified that

$$z_0(u)^- = -\frac{1}{2u} - \frac{1}{8u^4} - \frac{5}{32u^7} - \frac{11}{32u^{10}} - \frac{539}{512u^{13}} - \cdots \tag{12.4}$$

[see (9.11)].

The asymptotic expansion of $v_0(u)$ for $u \to +\infty$ is obtained directly from Eq. (9.1). Like any solution of this equation bounded on the whole positive half-line, the function $v_0(u)$

*The asymptotic expansion of $F(x)$, $-\infty < x < \infty$, for $x \to -\infty$, will be denoted by $F(x)^-$, and the expansion for $x \to +\infty$ will be denoted by $F(x)^+$.

has the asymptotic expansion*

$$v_0(u)^+ = \sum_{k=1}^{\infty}{}^* \frac{b_{k,0}^0}{u^k} + b_{0,0}^0. \tag{12.5}$$

The coefficients $b_{k,0}^0$, $k \neq 0$ can be successively calculated from the recurrence formula

$$b_{1,0}^0 = -1, \tag{12.6}$$
$$b_{k,0}^0 = -\frac{k-2}{k}\left(b_{0,0}^0 b_{k-2,0}^0 + \frac{1}{2}\sum_{\nu=1}^{k-3} b_{\nu,0}^0 b_{k-2-\nu,0}^0\right), \quad k \geqslant 2.$$

The constant $b_{0,0}^0$ is found by using the specific definition of the solution $v_0(u)$, selecting this function from the set of all solutions of (9.1) bounded for $u \to +\infty$, namely,

$$b_{0,0}^0 = \Omega_0 \tag{12.7}$$

[see (9.9)].

It is easily verified, in particular, that

$$v_0(u)^+ = \Omega_0 - \frac{1}{u} + \frac{\Omega_0}{3u^3} - \frac{1}{4u^4} - \frac{\Omega_0^2}{5u^5} + \frac{7\Omega_0}{18u^6} + \frac{4\Omega_0^3 - 5}{28u^7} + \ldots \tag{12.8}$$

[cf. (9.12)].

We now consider the functions $v_n(u)$, $n = 1,2,\ldots$, introduced in Sec. 10. The function $v_n(u)$ has the following asymptotic expansion for $u \to -\infty$:**

$$v_n(u)^- = u^{n-1}\sum_{k=0}^{\infty}\frac{a_k^n}{u^{3k}}, \quad n \geqslant 1. \tag{12.9}$$

---

*Series (sums) in which summation is over all values of an index except that following the first value will be indicated by Σ^*; for example,

$$\sum_{k=m}^{\infty}{}^* c_k \equiv \sum_{\substack{k=m,\\ k\neq m+1}}^{\infty} c_k = c_m + c_{m+2} + c_{m+3} + \ldots$$

**For $n = 0$, (12.9) yields the asymptotic expansion (12.2) of $z_0(u)$ for $u \to -\infty$.

There is a recurrence relation yielding the coefficients in this series; for example, it is easily proved that

$$
\begin{aligned}
v_1(u)^- &= -\frac{\tilde{\gamma}'_\xi}{2} - \frac{\tilde{\gamma}'_\xi}{8u^3} + \ldots, \\
v_2(u)^- &= \frac{2\tilde{\gamma}'_\eta - \tilde{\gamma}^{(2)}_\xi}{4} u + \frac{\tilde{\gamma}'_\eta}{4u^2} + \ldots, \qquad (12.10) \\
v_3(u)^- &= \frac{6\tilde{\gamma}''_{\xi\eta} - \tilde{\gamma}^{(3)}_\xi}{12} u^2 + \frac{\tilde{\gamma}^{(3)}_\xi + 6\gamma''_{\eta\xi} + 6\tilde{\gamma}'_\xi \tilde{\gamma}'_\eta + 3\tilde{\gamma}'_\xi \tilde{\gamma}^{(2)}_\xi}{48u} + \ldots
\end{aligned}
$$

[cf (11.1)].

There is a very important relation among the coefficients in the asymptotic expansions (12.2), (12.9), and (7.2):

$$a_k^n = \frac{1}{\gamma^{k+1}} \eta_n^{k+1}, \quad n \geqslant 0, \quad k \geqslant 0. \qquad (12.11)$$

This relation is shown schematically in the table in Fig. 38. The entries in the rows are the coefficients of the expansions of the functions $\gamma^{-i}\eta_i(\xi)$, $i = 1, 2, \ldots$, in powers of ξ for $\xi \to -0$. The columns give the coefficients of expansions of the functions $z_0(u)$, $v_m(u)$, $m = 1, 2, \ldots$, in the same powers of u for $u \to -\infty$ (the numbers in parentheses are the powers).

	$z_0(u)$	$v_1(u)$	$v_2(u)$	...	$v_m(u)$	...
$\frac{1}{\gamma}\eta_1(\xi)$	$\frac{\eta_0^1}{\gamma} = a_0^0$ (−1)	$\frac{\eta_1^1}{\gamma} = a_0^1$ (0)	$\frac{\eta_2^1}{\gamma} = a_0^2$ (1)	...	$\frac{\eta_m^1}{\gamma} = a_0^m$ $(m-1)$	...
$\frac{1}{\gamma^2}\eta_2(\xi)$	$\frac{\eta_0^2}{\gamma^2} = a_1^0$ (−4)	$\frac{\eta_1^2}{\gamma^2} = a_1^1$ (−3)	$\frac{\eta_2^2}{\gamma^2} = a_1^2$ (−2)	...	$\frac{\eta_m^2}{\gamma^2} = a_1^m$ $(m-4)$	...
$\frac{1}{\gamma^3}\eta_3(\xi)$	$\frac{\eta_0^3}{\gamma^3} = a_2^0$ (−7)	$\frac{\eta_1^3}{\gamma^3} = a_2^1$ (−6)	$\frac{\eta_2^3}{\gamma^3} = a_2^2$ (−5)	...	$\frac{\eta_m^3}{\gamma^3} = a_2^m$ $(m-7)$	...
...	...	...	...	...	...	...
$\frac{1}{\gamma^i}\eta_i(\xi)$	$\frac{\eta_0^i}{\gamma^i} = a_{i-1}^0$ $(-3i+2)$	$\frac{\eta_1^i}{\gamma^i} = a_{i-1}^1$ $(-3i+3)$	$\frac{\eta_2^i}{\gamma^i} = a_{i-1}^2$ $(-3i+4)$	...	$\frac{\eta_m^i}{\gamma^i} = a_{i-1}^m$ $(-3i+2+m)$	...
...	...	...	...	...	...	...

Fig. 38

The proof of (12.9) involves a more detailed study of the calculations used in the proof of Lemma 5. First, we see that the function $\mathcal{M}(u)$ introduced in (10.11) has the asymptotic expansion

$$\mathcal{M}(u)^{-} = \Re e^{-\frac{4}{3}u^3} u^2 \sum_{k=0}^{\infty} \frac{m_k^-}{u^{3k}}$$

[cf. (11.3)]. If we assume that the functions $v_1(u), \ldots, v_{n-1}(u)$ have expansions of the form (12.9), we can easily verify that

$$\sum_{\substack{i_1+\ldots+i_\nu=l,\\ i_j\geqslant 2}} v_{i_1-2} \ldots v_{i_\nu-2} = u^{l-3\nu} \sum_{k=0}^{\infty} \frac{a_k^{l,\nu}}{u^{3k}}, \qquad (12.12)$$

where v_0 should be understood to be the functions $z_0(u)$ with the asymptotic expansion (12.2). Now, by virtue of the definition (10.12) of $v_n(u)$, (10.8), (10.6), and (12.12), elementary calculations yield the asymptotic series for $\Phi_n(u)$ for $u \to -\infty$; we then obtain (12.9).

For $n = 0$ and all $k \geq 0$, (12.11) is obtained by comparing (7.3) with (12.3). To prove it for arbitrary $n \geq 1$, we must use general recurrence reations, i.e., we must substitute the coefficient η_n^{k+1} of ξ^{n-3k-1} in the expansion of $\eta_{k+1}(\xi)$ for $\xi \to -0$, starting from (6.10), and the coefficient a_k^n of u^{-3k} in the expansion of $v_n(u)$ for $u \to -\infty$, starting from (10.12), (10.11), (10.8), and (10.6).

There is no power series asymptotic expansion of $v_n(u)$ for $u \to +\infty$; a more complex structure must be used, with terms of the form $u^k \ln^\nu u$. The asymptotic expansion of $v_n(u)$ is

$$v_n(u)^{+} = \sum_{\nu=0}^{n} \sum_{k=3\nu+1-n}^{\infty}{}^{*} \frac{b_{k,\nu}^n \ln^\nu u}{u^k} + b_{0,\pi(n)}^n \ln^{\pi(n)} u, \quad n \geqslant 0, \qquad (12.13)$$

where

$$\pi(n) = \left[\frac{n}{3}\right] + \begin{cases} 0, & \text{for } n \not\equiv 1 \pmod 3, \\ 1, & \text{for } n \equiv 1 \pmod 3, \end{cases} \qquad (12.14)$$

is an integer-valued function of an integer-valued argument (its graph is shown in Fig. 39).* The coefficients in (12.13) can be calculated by using a very involved recurrence formula.

*For $n = 2$, (12.13) coincides with the asymptotic expansion (12.5) of $v_0(u)$ for $u \to +\infty$.

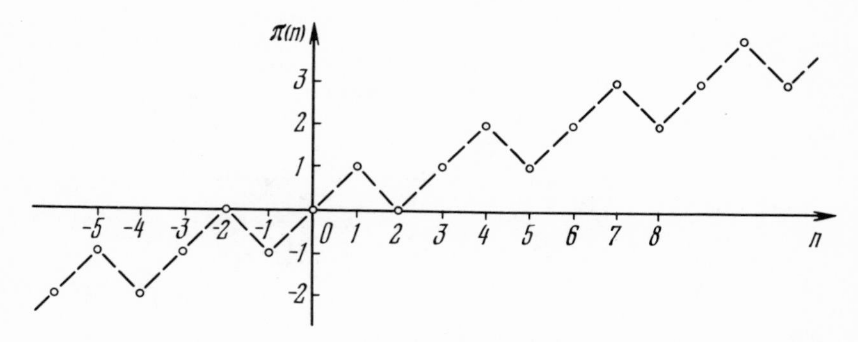

Fig. 39

It should be noted that this formula enables us to calculate the coefficients $b^n_{k,\nu}$ successively if $b^n_{0,0}$ is known. The constant

$$b^n_{0,\,0} = \Omega_n \tag{12.15}$$

characterizes the specific property of $v_n(u)$ distinguishing this function in the set of all solutions of the nth-order equation (10.9) [cf. (12.7)]. It is easily verified that

$$\begin{aligned} v_1(u)^+ &= \tilde{\gamma}'_\xi \ln u + \Omega_1 + \frac{\Omega_0 \tilde{\gamma}'_\xi}{2u^2} + \frac{\tilde{\gamma}'_\xi \ln u}{3u^3} + \frac{3\Omega_1 - 2\tilde{\gamma}'_\xi}{9u^3} + \ldots, \\ v_2(u)^+ &= \frac{\tilde{\gamma}^{(2)}_\xi}{2} u + \Omega_2 + \frac{\Omega_0 (\tilde{\gamma}^{(2)}_\xi - 2\tilde{\gamma}'_\eta)}{2u} + \\ &+ \frac{\tilde{\gamma}'^2_\xi \ln u}{2u^2} + \frac{2\Omega_1 \tilde{\gamma}'_\xi + 2\tilde{\gamma}'_\eta + \tilde{\gamma}'^2_\xi}{4u^2} + \ldots, \\ v_3(u)^+ &= \frac{\tilde{\gamma}^{(3)}_\xi}{12} u^2 + \frac{\Omega_0 (6\tilde{\gamma}''_{\xi\eta} - \gamma^{(3)}_\xi)}{6} \ln u + \Omega_3 + \frac{\tilde{\gamma}'_\xi (\tilde{\gamma}^{(2)}_\xi - 2\tilde{\gamma}'_\eta) \ln u}{2u} + \ldots \end{aligned} \tag{12.16}$$

[cf. (12.7)].

To prove (12.13), it is sufficient to consider in more detail the calculation used to establish Lemma 5. First, $\mathcal{M}(u)$ has the asymptotic expansion

$$\mathcal{M}(u)^+ = 1 + \sum_{k=3}^{\infty}{}^* \frac{m_k^+}{u^k}$$

[cf. (11.4)]. Assuming that the functions $v_n(u), \ldots, v_{n-1}(u)$ has asymptotic expansions of the form (12.13), we can apply elementary transformations to find the asymptotic series for

$u \to +\infty$ for the expression

$$\sum_{\substack{i_1+\ldots+i_\nu=l,\\ i_j\geq 0}} v_{i_1}(u)\ \ldots\ v_{i_\nu}(u),$$

and then, by employing (10.8) and (10.6), obtain the expansion

$$\Phi_n(u)^+ = \sum_{\nu=0}^{n} \sum_{k=3\nu+2-n}^{\infty}{}^{*} \frac{\tilde{b}^n_{k,\nu}\ln^\nu u}{u^k}.$$

Formula (12.13) now follows directly from the definition (10.12) of $v_n(u)$.

We now discuss an important fact. It is clear that the $v_n(u)$, $n \geq 1$, depend on the specific form of the system (8.4) or, equivalently, on Eq. (10.2). Moreover, the nature of this dependence can easily be determined by analyzing the structure of formulas (10.6). In fact, each function $v_n(u)$, $n \geq 1$, is a linear combination of a certain number of universal functions [not depending on the specific nature of Eq. (10.2)] with coefficients that can be expressed in terms of the values of the function $\gamma(\xi,\eta)$ and some of its derivatives at the origin. For example, it can be shown that

$$\begin{aligned} v_1(u) &= \tilde{\gamma}'_\xi \boldsymbol{v}_1(u), \\ \boldsymbol{v}_1(u) &= \mathcal{M}(u)\int_{-\infty}^{u} \theta \frac{dv_0}{d\theta}\frac{d\theta}{\mathcal{M}(\theta)}; \\ v_2(u) &= \frac{\tilde{\gamma}_\xi^{(2)}}{2}\boldsymbol{v}_{2,1}(u) + \tilde{\gamma}'_\eta \boldsymbol{v}_{2,2}(u) + \tilde{\gamma}'^2_\xi \boldsymbol{v}_{2,3}(u), \end{aligned} \qquad (12.17)$$

where

$$\boldsymbol{v}_{2,1}(u) = \mathcal{M}(u)\int_{-\infty}^{u} \theta^2 \frac{dv_0}{d\theta}\frac{d\theta}{\mathcal{M}(\theta)},$$

$$\boldsymbol{v}_{2,2}(u) = \mathcal{M}(u)\int_{-\infty}^{u} v_0(\theta) \frac{dv_0}{d\theta}\frac{d\theta}{\mathcal{M}(\theta)},$$

$$\boldsymbol{v}_{2,3}(u) = \mathcal{M}(u)\int_{-\infty}^{u} \boldsymbol{v}_1(\theta)\left(\frac{dv_0}{d\theta}\right)^2 \left[\boldsymbol{v}_1(\theta)\frac{dv_0}{d\theta} - \theta\right]\frac{d\theta}{\mathcal{M}(\theta)}.$$

Hence, to find the coefficients in the series (10.4), we need only know the values of $\gamma(\xi,\eta)$ and its derivatives at the origin, and there is no need to solve the nondegenerate system (8.4). There is a similar result for the constant (12.15).

13. Regularization of Improper Integrals

We shall consider integrals with variable limits of integration in a certain interval (a segment of a half-line), which are singular at the endpoint, so that the integrand, in general, at both ends of the interval, is not integrable. To find the asymptotic properties of these integrals as the variable limits approach the ends of the interval, we introduce a special form of generalized integral; in fact we obtain a certain regularization of improper integrals. The regularization method used below is based on a method of dealing with nonintegrable singularities of integrands, introduced by Hadamard [1].

Let $h(x) \in C^{\infty}[0,p]$, where $p > 0$ is a fixed number, let $h(0) \neq 0$, and let n be a positive integer. We shall obtain an asymptotic expansion for $x \to +0$ of the function

$$H(x) = \int_x^p \frac{h(\xi)}{\xi^n} d\xi \tag{13.1}$$

that is clearly defined if $0 < x \leq p$. Subtracting and adding the first n terms of the Maclaurin series for $h(\xi)$, we obtain

$$\begin{aligned} H(x) &= \int_x^p \frac{1}{\xi^n}\left[h(\xi) - \sum_{k=0}^{n-1} \frac{h^{(k)}(0)}{k!}\xi^k\right] d\xi + \\ &+ \int_x^p \sum_{k=0}^{n-1} \frac{h^{(k)}(0)}{k!}\frac{d\xi}{\xi^{n-k}} = \int_0^p \frac{1}{\xi^n}\left[h(\xi) - \sum_{k=0}^{n-1}\frac{h^{(k)}(0)}{k!}\xi^k\right] d\xi - \\ &- \int_0^x \frac{1}{\xi^n}\left[h(\xi) - \sum_{k=0}^{n-1}\frac{h^{(k)}(0)}{k!}\xi^k\right] d\xi - \sum_{k=0}^{n-2}\frac{h^{(k)}(0)}{k!\,(n-k-1)\,p^{n-k-1}} + \\ &+ \frac{h^{(n-1)}(0)}{(n-1)!}\ln p + \sum_{k=0}^{n-2}\frac{h^{(k)}(0)}{k!\,(n-k-1)\,x^{n-k-1}} - \frac{h^{(n-1)}(0)}{(n-1)!}\ln x. \end{aligned}$$

The first integral exists, and the asymptotic expansion of the second for $x \to +0$ is easily obtained. Thus, for $x \to +0$, the function (13.1) has the following asymptotic representation:

$$H(x) = -\sum_{\substack{k=1-n,\\ k\neq 0}}^{\infty} \frac{h^{(n+k-1)}(0)}{(n+k-1)!\,k} x^k - \frac{h^{(n-1)}(0)}{(n-1)!}\ln x + \oint_0^p \frac{h(\xi)}{\xi^n} d\xi, \quad x \to +0, \tag{13.2}$$

where

$$\oint_0^p \frac{h(x)}{x^n}\,dx = \int_0^p \left[\frac{h(x)}{x^n} - \frac{h(0)}{x^n} - \frac{h'(0)}{1!\,x^{n-1}} - \cdots - \frac{h^{(n-1)}(0)}{(n-1)!\,x}\right] dx -$$
$$- \frac{h(0)}{(n-1)\,p^{n-1}} - \frac{h'(0)}{1!\,(n-2)\,p^{n-2}} - \cdots - \frac{h^{(n-2)}(0)}{(n-2)!\,p} + \frac{h^{(n-1)}(0)}{(n-1)!}\ln p \qquad (13.3)$$

is a completely determined constant.

Relation (13.3) is the definition of a special *generalized integral* of a function $x^{-n}h(x)$, not integrable in the usual sense on the interval $0 \leq x \leq p$.

We now discuss some properties of the generalized integral (13.3). It is clearly a linear functional:

$$\oint_0^p \left[\alpha_1 \frac{h_1(x)}{x^n} + \alpha_2 \frac{h_2(x)}{x^m}\right] dx = \alpha_1 \oint_0^p \frac{h_1(x)}{x^n}\,dx + \alpha_2 \oint_0^p \frac{h_2(x)}{x^m}\,dx. \qquad (13.4)$$

It can be verified directly that

$$\oint_0^p \frac{h(x)}{x^n}\,dx + \oint_p^{p^*} \frac{h(x)}{x^n}\,dx = \oint_0^{p^*} \frac{h(x)}{x^n}\,dx, \quad p > 0, \quad p^* > 0. \qquad (13.5)$$

If we consider a generalized integral in the sense (13.3) with a variable upper limit, the definition yields the following analog of the Newton-Leibnitz formula:

$$\frac{d}{dx}\oint_0^x \frac{h(\xi)}{\xi^n}\,d\xi = \frac{h(x)}{x^n}, \quad 0 < x \leqslant p. \qquad (13.6)$$

Formula (13.6) enables us to use generalized integrals to obtain solutions of certain ordinary differential equations with singular right-hand sides. For example, the solution of the equation

$$\frac{dy}{dx} = \frac{h(x)}{x^n}, \quad h(x) \in C^\infty[0, p], \quad h(0) \neq 0, \qquad (13.7)$$

is

$$y = C + \oint_0^x \frac{h(\xi)}{\xi^n}\,d\xi, \quad 0 < x \leqslant p, \qquad (13.8)$$

where C is an arbitrary constant. The function (13.8) is not defined for $x = 0$ and increases without bound in absolute value when $x \to +0$; its asymptotic properties for $x \to +0$ are described by (13.3).

The idea of generalized integrals can be extended to apply to more general functions. Suppose that $f(x) \in C^\infty[0,p]$, $g(x) \in C^\infty[0,p]$, $f(0) \neq 0$, and $g(x) \equiv x^n g_1(x)$, where $g_1(x) \neq 0$ for $0 \leq x \leq p$. It is natural to introduce the following definition:

$$\oint_0^p \frac{f(x)}{g(x)}\, dx = \oint_0^p \frac{h(x)}{x^n}\, dx, \text{ where } h(x) \equiv \frac{f(x)}{g_1(x)}. \tag{13.9}$$

The rule for transforming a generalized integral (13.3) by the infinitely differentiable change

$$x = \varphi(y); \quad \varphi(0) = 0, \quad \varphi(q) = p; \quad \varphi'(y) > 0 \text{ for } 0 \leqslant y \leqslant q, \tag{13.10}$$

of the variable of integration yields

$$\oint_0^p \frac{h(x)}{x^n}\, dx = \oint_0^q \frac{h(\varphi(y))\,\varphi'(y)}{\varphi^n(y)}\, dy + \sum_{k=1}^{n} \frac{h^{(n-k)}(0)}{(n-k)!}\, \Phi_k(\varphi'(0), \ldots, \varphi^{(k)}(0)). \tag{13.11}$$

The integral on the right in (13.11) is understood in the sense of (13.9), and the $\Phi_k(z_1, \ldots, z_k)$, $k = 1, \ldots, n$, are certain universal functions whose form is independent of the function $h(x)$, n, p, and the substitution (13.10):

$$\begin{aligned} \Phi_1(z_1) &= \ln|z_1|, \\ \Phi_2(z_1, z_2) &= \frac{z_2}{2z_1^2}, \\ \Phi_3(z_1, z_2, z_3) &= \frac{4z_1 z_3 - 9z_2^2}{24 z_1^4}, \\ &\ldots\ldots\ldots\ldots \end{aligned} \tag{13.12}$$

We outline the proof of (13.11). Using (13.3), we make the substitution (13.10) in the corresponding classical integral:

$$\oint_0^p \frac{h(x)}{x^n}\, dx = \oint_0^q \frac{1}{\varphi^n(y)} \left[h(\varphi(y)) - \sum_{k=0}^{n-1} \frac{h^{(k)}(0)}{k!} \varphi^k(y) \right] \varphi'(y)\, dy -$$

$$- \sum_{k=0}^{n-2} \frac{h^{(k)}(0)}{k!\,(n-k-1)\,p^{n-k-1}} + \frac{h^{(n-1)}(0)}{(n-1)!} \ln p.$$

On the other hand, (13.9) implies that

$$\oint_0^q \frac{h(\varphi(y))\,\varphi'(y)}{\varphi^n(y)}\,dy = \oint_0^q \frac{\hbar(y)}{y^n}\,dy =$$

$$= \oint_0^q \frac{1}{y^n}\left[\hbar(y) - \sum_{k=0}^{n-1} \frac{\hbar^{(k)}(0)}{k!}\,y^k\right]dy - \sum_{k=0}^{n-2} \frac{\hbar^{(k)}(0)}{k!\,(n-k-1)\,q^{n-k-1}} + \frac{\hbar^{(n-1)}(0)}{(n-1)!}\ln q,$$

where

$$\hbar(y) = \frac{y^n h(\varphi(y))\,\varphi'(y)}{\varphi^n(y)}. \tag{13.13}$$

Hence

$$\oint_0^p \frac{h(x)}{x^n}\,dx = \oint_0^q \frac{\hbar(y)}{y^n}\,dy + \lim_{\delta\to 0}\int_\delta^q \left[\frac{1}{y^n}\sum_{k=0}^{n-1}\frac{\hbar^{(k)}(0)}{k!}\,y^k - \frac{\varphi'(y)}{\varphi^n(y)}\sum_{k=0}^{n-1}\frac{h^{(k)}(0)}{k!}\,\varphi^k(y)\right]dy +$$

$$+ \sum_{k=0}^{n-2}\frac{1}{k!\,(n-k-1)}\left[\frac{\hbar^{(k)}(0)}{q^{n-k-1}} - \frac{h^{(k)}(0)}{p^{n-k-1}}\right] - \frac{1}{(n-1)!}\left[\hbar^{(n-1)}(0)\ln q - h^{(n-1)}(0)\ln p\right] =$$

$$= \oint_0^q \frac{h(\varphi(y))\,\varphi'(y)}{\varphi^n(y)}\,dy + \lim_{\delta\to 0}\left\{\sum_{k=0}^{n-2}\frac{1}{k!\,(n-k-1)}\left[\frac{\hbar^{(k)}(0)}{\delta^{n-k-1}} - \frac{h^{(k)}(0)}{\varphi^{n-k-1}(\delta)}\right] -\right.$$

$$\left. - \frac{1}{(n-1)!}\left[\hbar^{(n-1)}(0)\ln\delta - h^{(n-1)}(0)\ln\varphi(\delta)\right]\right\}.$$

The relations

$$\hbar(0) = \frac{h(0)}{[\varphi'(0)]^{n-1}},$$

$$\hbar'(0) = \frac{h'(0)}{[\varphi'(0)]^{n-2}} - (n-2)\frac{\varphi''(0)}{2\,[\varphi'(0)]^n}\,h(0),$$

$$\cdots\cdots\cdots\cdots\cdots\cdots\cdots\cdots$$

$$\hbar^{(n-1)}(0) = h^{(n-1)}(0)$$

will be needed later; they are proved by successive differentiation of (13.13). These relations show that the limit obtained in the above analysis actually exists and is equal to the term outside the integral in (13.11). The general relation between $\hbar^{(k)}(0)$ and $h(0)$, $h'(0),\ldots,h^{(k)}(0)$, and expressions for the functions $\Phi_k(z_1,\ldots,z_k)$, $k = 1,\ldots,n$, can be obtained by employing the formula for the kth derivative of a composite function [16].

If $h(x) \in C^{\infty}[-p,0]$, where $p > 0$ and $h(0) \neq 0$, then by analogy with (13.3), we define the following:

$$\oint_{-p}^{0} \frac{h(x)}{x^n} dx = \int_{-p}^{0} \left[\frac{h(x)}{x^n} - \frac{h(0)}{x^n} - \frac{h'(0)}{1!\,x^{n-1}} - \dots - \frac{h^{(n-1)}(0)}{(n-1)!\,x}\right] dx +$$

$$+ \frac{(-1)^{n-1} h(0)}{(n-1)\,p^{n-1}} + \frac{(-1)^{n-2} h'(0)}{1!\,(n-2)\,p^{n-2}} + \dots + \frac{(-1)\,h^{(n-2)}(0)}{(n-2)!\,p} - \frac{h^{(n-1)}(0)}{(n-1)!} \ln p. \quad (13.14)$$

By using this generalized integral we can easily obtain the asymptotic expansion of the function

$$H_1(x) = \int_{-p}^{x} \frac{h(\xi)}{\xi^n} d\xi$$

for $x \to -0$ [similar to the expansion (13.2)]. The integrals

$$\oint_{\alpha}^{\beta} \frac{h(x)}{(x-\alpha)^n} dx, \qquad \oint_{\alpha}^{\beta} \frac{h(x)}{(x-\beta)^n} dx,$$

are defined in the obvious fashion if $h(x) \in C^{\infty}[\alpha\ \beta]$, and $h(\alpha) \neq 0$ or $h(\beta) \neq 0$, respectively. This enables us to apply an infinitely differentiable change of the variable of integration of more general form than (13.10) to the generalized integral (13.3) [or (13.4)]. Thus the rule for transforming the integral (13.3) by the variable change

$$x = \varphi(y); \quad \varphi(\alpha) = 0, \quad \varphi(\beta) = p; \quad \varphi'(y) \neq 0 \quad (13.15)$$

is given by the formula

$$\oint_{0}^{p} \frac{h(x)}{x^n} dx = \oint_{\alpha}^{\beta} \frac{h(\varphi(y))\,\varphi'(y)}{\varphi^n(y)} dy + \sum_{k=1}^{n} \frac{h^{(n-k)}(0)}{(n-k)!} \Phi_k(\varphi'(\alpha), \dots, \varphi^{(k)}(\alpha)), \quad (13.16)$$

where $\Phi_k(z_1,\dots,z_k)$, $k = 1,\dots,n$ are the same functions as in (13.11) [see (13.12)].

The method we have used for defining generalized integrals, by the subtraction of the nonintegrable singularities of the integrand, can also be applied to a wider class of functions possessing singularities more general than power singularities. For example, if $h(x) \in C^{\infty}[0,p]$ with $p > 0$, $h(0) \neq 0$, and n and m are positive integers, then by definition

$$\oint_0^p \frac{\ln^m(x)}{x^n} h(x)\,dx = \int_0^p \frac{\ln^m(x)}{x^n}\left[h(x) - \sum_{\nu=0}^{n-1} \frac{h^{(\nu)}(0)}{\nu!} x^\nu\right] dx -$$

$$-\sum_{\nu=0}^{n-2}\sum_{k=1}^{m} \frac{m!\,h^{(\nu)}(0)\ln^k p}{\nu!\,k!\,(n-\nu-1)^{m-k+1}\,p^{n-\nu-1}} + \frac{h^{(n-1)}(0)\ln^{m+1} p}{(n-1)!\,(m+1)}. \qquad (13.17)$$

The properties of this generalized integral are similar to the properties of the integral (13.3).

Finally, we can also introduce a natural definition of a contour integral

$$\oint_\Gamma f(x, y)\,dx + g(x, y)\,dy$$

when $f(x,y)$ and $g(x,y)$ are infinitely differentiable on Γ everywhere except at a finite number of points at which the functions have power or power-logarithmic singularities.

Up to this stage, we have considered generalized integrals for the case of a finite range of integration with a singularity at one end point. We shall show how the same approach (subtraction of the nonintegrable singularities of the integrand) can be used in many case to define generalized integrals on a half-line (with the singularity at infinity).

Let $h(x) \in C[0,\infty)$, for large positive values of the argument, have the asymptotic expansion

$$h(x) = a_n x^n + a_{n-1}x^{n-1} + \dots + a_1 x + a_0 + \frac{a_{-1}}{x} + \frac{a_{-2}}{x^2} + \dots =$$

$$= \sum_{k=-n}^{\infty} \frac{a_{-k}}{x^k}, \quad x \to +\infty. \qquad (13.18)$$

We shall find the asymptotic expansion, for $x \to +\infty$, of the function

$$\tilde{H}(x) = \int_0^x h(\xi)\,d\xi, \qquad (13.19)$$

which is defined for $0 \leq x \leq \infty$. Putting

$$\Delta^+(x) = \begin{cases} 0 & \text{for } 0 \leqslant x < 1, \\ 1 & \text{for } 1 \leqslant x < \infty, \end{cases}$$

we write the function (13.19) in the form

$$\tilde{H}(x)=\int_0^x\left[h(\xi)-\sum_{k=-n}^{0}\frac{a_{-k}}{\xi^k}-\frac{a_{-1}\Delta^+(\xi)}{\xi}\right]d\xi+\int_0^x\left[\sum_{k=-n}^{0}\frac{a_{-k}}{\xi^k}+\frac{a_{-1}\Delta^+(\xi)}{\xi}\right]d\xi=$$

$$=\int_0^\infty\left[h(\xi)-\sum_{k=0}^{n}a_k\xi^k-\frac{a_{-1}\Delta^+(\xi)}{\xi}\right]d\xi-\int_x^\infty\left[h(\xi)-\sum_{k=0}^{n}a_k\xi^k-\frac{a_{-1}\Delta^+(\xi)}{\xi}\right]d\xi+$$

$$+\sum_{k=0}^{n}\frac{a_kx^{k+1}}{k+1}+a_{-1}\ln x.$$

Here the first integral exists, and the asymptotic expansion of the second for $x\to+\infty$ is easily found. Hence, for $x\to+\infty$, the function (13.19) has the asymptotic expansion

$$\tilde{H}(x)=\sum_{k=0}^{n}\frac{a_kx^{k+1}}{k+1}+a_{-1}\ln x+$$
$$+\oint_0^\infty h(\xi)\,d\xi-\sum_{k=2}^{\infty}\frac{a_{-k}}{(k-1)x^{k-1}},\quad x\to+\infty, \tag{13.20}$$

where

$$\oint_0^\infty h(x)\,dx=\int_0^\infty\left[h(x)-a_nx^n-\ldots-a_1x-a_0-\frac{a_{-1}\Delta^+(x)}{x}\right]dx \tag{13.21}$$

is a definite constant.

Relation (13.21) defines the *generalized integral* over $0\le x<\infty$ of the function with the asymptotic power series (13.18) for $x\to+\infty$. For the half-line $-\infty<x\le 0$, the function $\Delta^+(x)$ is replaced by the function

$$\Delta^-(x)=\begin{cases}1 & \text{for } -\infty<x\leqslant-1,\\ 0 & \text{for } -1<x\leqslant 0.\end{cases}$$

Generalized integrals on $0\le x<\infty$ of functions whose asymptotic expansions are of the power-logarithmic form for $x\to+\infty$ are defined similarly [see (12.13)].

14. Asymptotic Expansions for the End of a Junction Part of a Trajectory

As we have already shown [see (11.1) and (11.2)], the functions $v_i(u)$ increase unboundedly in absolute value when $u \to -\infty$ (for $i = 2, 3, \ldots$) and when $u \to +\infty$ (for $i = 1, 2, \ldots$). It is thus not possible to use the series in (10.4) to obtain asymptotic representations of $\mathfrak{T}_\varepsilon$ on the whole junction part, i.e., for $-q \le \xi \le q$. To find asymptotic approximations for this part, we must replace the power expansion by a more complicated expansion using terms $\mu^n \ln^\nu (1/\mu)$.

In Sec. 6, we indicated that the equation of the junction part of the trajectory $\mathfrak{T}_\varepsilon$ can be written in the form (6.1), and the function involved is a solution of Eq. (6.2). Here, instead of ε we use the parameter μ [see (8.2)]; then the function (6.1) will be denoted by

$$\eta = \overset{*}{\eta}_{\mathfrak{T}}(\xi, \mu) \equiv \eta_{\mathfrak{T}}(\xi, \mu^3/\gamma), \quad -q \leqslant \xi \leqslant q, \tag{14.1}$$

and the equation satisfied by the function (14.1) can be written as

$$\frac{d\eta}{d\xi} = \mu^3 \frac{\tilde{\gamma}(\xi, \eta)}{\xi^2 + \eta}. \tag{14.2}$$

We shall use (14.2) to find the asymptotic expansion for the trajectory $\mathfrak{T}_\varepsilon$ at *the end of the junction part* on $\sigma_2 \le \xi \le q$, where $\sigma_2 > 0$ tends to zero with ε. In other words, we consider asymptotic expansions of the function

$$\eta = \overset{*}{\eta}_{\mathfrak{T}}(\xi, \mu), \quad \sigma_2 \leqslant \xi \leqslant q, \tag{14.3}$$

coinciding with the function (14.1) [and hence with the function (6.1)] on their common interval of definition.

We construct the formal series

$$\eta = \sum_{n=2}^{\infty} \mu^n \sum_{\nu=0}^{\pi(n-2)} \zeta_{n,\nu}(\xi) \ln^\nu \frac{1}{\mu} \tag{14.4}$$

so that it *formally* satisfies (14.2) for $0 < \xi \le q$. We shall explain in Sec. 16 why we use a series of this form; we use it to obtain asymptotic approximations for the values $\overset{*}{\eta}_{\mathfrak{T}}(\sigma_2, \mu)$ for $\mathfrak{T}_\varepsilon$ at $\xi = \sigma_2$, calculated from formulas in Sec. 10.

The coefficients in this series are calculated as follows. We use the series (14.4) in (14.2), and expand the right side in a formal series of terms $\mu^n \ln^\nu(1/\mu)$. It is convenient to use the formulas

$$\frac{\tilde{\gamma}\left(\xi, \sum_{n=1}^{\infty} \mu^n z_n\right)}{\xi^2 + \sum_{n=1}^{\infty} \mu^n z_n} = \frac{\tilde{\gamma}(\xi, 0)}{\xi^2} +$$

$$+ \sum_{n=1}^{\infty} \mu^n \sum_{k=1}^{n} \sum_{m=0}^{k} \frac{(-1)^m \tilde{\gamma}_\eta^{(k-m)}(\xi, 0)}{(k-m)!\, \xi^{2m+2}} \sum_{\substack{i_1+\ldots+i_k=n,\\ i_j \geqslant 1}} z_{i_1} \ldots z_{i_k}, \tag{14.5}$$

which follow from (3.7) and (6.7). Employing the abbreviations

$$w_1 = 0, \quad w_n(\xi) = \sum_{\nu=0}^{\pi(n-2)} \zeta_{n,\nu}(\xi) \ln^\nu \frac{1}{\mu}, \quad n \geqslant 2, \tag{14.6}$$

and (14.5), we equate coefficients of powers of μ in the left and right sides of (14.2) to obtain

$$w_2'(\xi) = 0, \quad w_3'(\xi) = \frac{\tilde{\gamma}(\xi, 0)}{\xi^2}, \quad w_4'(\xi) = 0,$$

$$w_n'(\xi) = \sum_{k=1}^{\left[\frac{n-3}{2}\right]} \sum_{m=0}^{k} \frac{(-1)^m \tilde{\gamma}_\eta^{(k-m)}(\xi, 0)}{(k-m)!\, \xi^{2m+2}} \sum_{\substack{i_1+\ldots+i_k=n-3,\\ i_j \geqslant 2}} w_{i_1} \ldots w_{i_k}, \; n \geqslant 5. \tag{14.7}$$

The function (12.14) satisfies the relation

$$\max_{\substack{i_1+\ldots+i_k=n-3,\\ i_j \geqslant 2,\, n \geqslant 5}} \{\pi(i_1-2) + \ldots + \pi(i_k-2)\} = \pi(n-2) - 1; \tag{14.8}$$

hence [see (14.6)]

$$\sum_{\substack{i_1+\ldots+i_k=n-3,\\ i_j \geqslant 2}} w_{i_1} \ldots w_{i_k} = \sum_{\nu=0}^{\pi(n-2)-1} W_{n,k}^{\nu}(\xi) \ln^\nu \frac{1}{\mu}, \quad n \geqslant 5; \tag{14.9}$$

for $n \geq 5$ the functions $W_{n,k}^{\nu}(\xi)$ can be expressed in terms of the coefficients $\zeta_{m,\nu}(\xi)$, $m \leq n-3$. It is clear from (14.9) that the right-hand side of the nth relation (14.7) with $n \geq 5$ is a polynomial in $\ln(1/\mu)$ of degree $\pi(n-2) - 1$ with coefficients independent of the functions $\zeta_{n,\nu}(\xi)$, $\nu = 0, 1, \ldots, \pi(n-2)$. By equating coefficients of powers of $\ln(1/\mu)$ in (14.7), we obtain the following relations:

$$\zeta'_{2,0}(\xi)=0; \quad \zeta'_{3,1}(\xi)=0, \quad \zeta'_{3,0}(\xi)=\frac{\tilde{\gamma}(\xi,0)}{\xi^2};$$
$$\zeta'_{4,0}(\xi)=0; \quad \zeta'_{n,\pi(n-2)}(\xi)=0, \quad n\geqslant 5,$$
$$\zeta'_{n,\nu}(\xi)=\sum_{k=1}^{\left[\frac{n-3}{2}\right]}\sum_{m=0}^{k}\frac{(-1)^m\tilde{\gamma}_\eta^{(k-m)}(\xi,0)}{(k-m)!\,\xi^{2m+2}}W^{\nu}_{n,k}(\xi), \tag{14.10}$$
$$n\geqslant 5, \quad \nu=0, 1, \ldots, \pi(n-2)-1.$$

These are differential equations with separable variables but, in general, with a singularity in the right sides for $\xi = 0$ [equations of the type (13.7), for example]. We note that none of the equations (14.10) has an extra condition to distinguish a definite solution. Hence Eqs. (14.10) successively determine the coefficients $\zeta_{n,\nu}(\xi)$, each to within an arbitrary additive constant [see (13.8)]. We select the $\zeta_{n,\nu}(\xi)$ as follows:

$$\zeta_{2,0}(\xi)=b^0_{0,0};$$
$$\zeta_{3,1}(\xi)=b^1_{0,1}, \quad \zeta_{3,0}(\xi)=b^1_{0,0}+\oint_0^{\xi}\frac{\tilde{\gamma}(\theta,0)}{\theta^2}\,d\theta;$$
$$\zeta_{4,0}(\xi)=b^2_{0,0}; \tag{14.11}$$
$$\zeta_{n,\pi(n-2)}(\xi)=b^{n-2}_{0,\pi(n-2)}, \quad n\geqslant 5,$$
$$\zeta_{n,\nu}(\xi)=b^{n-2}_{0,\nu}+\oint_0^{\xi}\sum_{k=1}^{\left[\frac{n-3}{2}\right]}\sum_{m=0}^{k}\frac{(-1)^m\tilde{\gamma}_\eta^{(k-m)}(\theta,0)}{(k-m)!\,\theta^{2m+2}}W^{\nu}_{n,k}(\theta)\,d\theta,$$
$$n\geqslant 5, \quad \nu=0, 1, \ldots, \pi(n-2)-1;$$

the constants $b^n_{0,\nu}$ occur in the asymptotic expansions (12.5), (12.13), and the $W^{\nu}_{n,k}(\xi)$ are determined successively by (14.9). The reasons for choosing these coefficients $\zeta_{n,\nu}(\xi)$ are discussed in Sec. 16.

It easily follows from the results in Sec. 13 that the $\zeta_{n,\nu}(\xi)$ are defined and smooth for $0 < \xi \leq q$. We stress that each of these functions can be effectively calculated in terms of values of $\gamma(\xi,\eta)$ and some of its derivatives on the part $0 < \xi \leq q$ of the curve (5.9).

In Secs. 15 and 16 we shall see that the partial sum

$$Z_n(\xi, \mu) = \sum_{k=2}^{n} \mu^k \sum_{\nu=0}^{\pi(k-2)} \zeta_{k,\nu}(\xi) \ln^\nu \frac{1}{\mu} \qquad (14.12)$$

of the series (14.4) is, for $\varepsilon \to 0$, the asymptotic expansion for the part (14.3) of the trajectory $\mathfrak{T}_\varepsilon$:

$$\eta^*_{\mathfrak{T}}(\xi, \mu) = \begin{cases} O(\varepsilon^{2/3}), \\ Z_2(\xi, \mu) + O(\varepsilon^{1-\lambda_2}), \\ Z_3(\xi, \mu) + O\left(\varepsilon^{\min\left(\frac{4}{3},\, \frac{5}{3} - 3\lambda_2\right)}\right), \\ Z_{N_4+3}(\xi, \mu) + O\left(\varepsilon^{\frac{N_4+4}{3} - \lambda_2(N_4+2)}\right), \end{cases} \quad \sigma_2 \leqslant \xi \leqslant q, \qquad (14.13)$$

where N_4 is an arbitrary positive integer and

$$\sigma_2 = \varepsilon^{\lambda_2}, \qquad 0 < \lambda_2 < 1/3. \qquad (14.14)$$

15. The Relation between Asymptotic Approximations and Actual Trajectories at the End of a Junction Part

Lemma 8. For $n = 2, 3, \ldots$ and $\nu = 0, 1, \ldots, \pi(n-2)$, there is a constant $M_{n,\nu} > 0$ such that

$$|\zeta_{n,\pi(n-2)}(\xi)| \leqslant M_{n,\pi(n-2)}, \quad 0 < \xi \leqslant q,$$

$$|\xi^{n-2-\nu}\zeta_{n,\nu}(\xi)| \leqslant M_{n,\nu}, \quad \nu = 0, 1, \ldots, \pi(n-2)-1, \quad 0 < \xi \leqslant q; \qquad (15.1)$$

$$|\xi^{n-1-\nu}\zeta'_{n,\nu}(\xi)| \leqslant M_{n,\nu}, \quad \nu = 0, 1, \ldots, \pi(n-2)-1, \quad 0 < \xi \leqslant q. \qquad (15.2)$$

We use (14.9)-(14.11), the properties of generalized integrals (Sec. 13), and the fact that $\gamma(\xi,\eta)$ and its derivatives are bounded in $\mathcal{U}$, and prove (15.2) by induction and then prove (15.1). It should be noted that all the functions $\zeta_{n,\pi(n-2)}(\xi)$, $n \geq 2$, are constants.

More detailed calculations show that, for $n \geq 0$ and $\nu = 0, 1, \ldots, \pi(n-2)$, the functions $\zeta_{n,\nu}(\xi)$ have the following asymptotic expansion:

$$\zeta_{n,\nu}(\xi) = \sum_{\varkappa=0}^{\pi(n-2)-\nu-1} \sum_{k=2-n+\varkappa+\nu}^{\infty} \zeta^{n,\nu}_{k,\varkappa} \xi^k \ln^\varkappa \xi + \zeta^{n,\nu}_{0,\pi(n-2)-\nu} \ln^{\pi(n-2)-\nu} \xi, \quad \xi \to +0, \qquad (15.3)$$

and there is a very cumbersome recurrence relation for the coefficients $\zeta^{n,\nu}_{k,\kappa}$. Since, in particular,

$$\zeta^{2,0}_{0,0}=b^0_{0,0}, \qquad \zeta^{3,0}_{-1,0}=-1,$$

$$\zeta^{n,0}_{2-n,0}=-\frac{n-4}{n-2}\left(\zeta^{2,0}_{0,0}\zeta^{n-2,0}_{4-n,0}+\frac{1}{2}\sum_{\varkappa=1}^{n-5}\zeta^{\varkappa+2,0}_{-\varkappa,0}\zeta^{n-\varkappa-2,0}_{\varkappa+4-n,0}\right), \qquad n\geqslant 4; \tag{15.4}$$

it is clear that the functions $\zeta_{n,0}(\xi)$, $n \geq 5$, tend either to $+\infty$ or $-\infty$ when $\xi \to +0$, depending on the value of n. In the general case, the functions $\zeta_{n,\nu}(\xi)$, $n \geq 5$, $\nu \neq \pi(n-2)$, also tend either to $+\infty$ or $-\infty$ when $\xi \to +0$, depending on the values of n and ν.

It is important that there is the following direct relation between the coefficients in the asymptotic series (15.3) and (12.13):

$$\zeta^{n,\nu}_{k,\varkappa}=C^{\nu}_{\nu+\varkappa}b^{n+k-2}_{-k,\nu+\varkappa},$$
$$n=3, 5, 6, \ldots, \nu=0, 1, \ldots, \pi(n-2)-1,$$
$$\varkappa=0, 1, \ldots, \pi(n-2)-1-\nu, \quad k\geqslant 2-n+\nu+\varkappa; \tag{15.5}$$
$$\zeta^{n,\nu}_{0,\pi(n-2)-\nu}=C^{\nu}_{\pi(n-2)}b^{n-2}_{0,\pi(n-2)},$$
$$n\geqslant 2, \qquad \nu=0, 1, \ldots, \pi(n-2).$$

This relation is shown schematically in Fig. 40. The points represent the coefficients $b^{n}_{k,\nu}$ of terms of the form $u^{-k}\ln^{\nu}u$ in asymptotic expansions, for $u \to +\infty$, of the functions $v_0(u)$, $v_1(u)$, $v_2(u)$,... . The broken lines pass through points representing coefficients in the asymptotic expansion, for $\xi \to +0$, of the corresponding function $\zeta_{m,\kappa}(\xi)$ as coefficients of terms $\xi^k \ln^{\nu}(\xi)$ [the binomial coefficients in formulas (15.5) are not shown]. To prove (15.5), we use general recurrence formulas [see (15.4) and (12.6)].

Lemma 9. Let

$$F_n(\xi, \mu, \sigma_2)=\frac{\mu^{n+1}}{\sigma_2^{n-1}}+\mu^{n+1}\sum_{\nu=0}^{\pi(n-1)-1}\int_{\sigma_2}^{\xi}\frac{d\theta}{\theta^{n-\nu}}\ln^{\nu}\frac{1}{\mu}, \quad n=2, 4, 5\ldots;$$

$$F_3(\xi, \mu, \sigma_2)=\mu^4+\frac{\mu^5}{\sigma_2^3}+\mu^5\int_{\sigma_2}^{\xi}\frac{d\theta}{\theta^4}.$$

For sufficiently small ε, there is a constant $C_n > 0$ such that the derivative (14.2) is positive at each point of the curve

$$\mathcal{K}_1(\xi, \eta)\equiv\eta-Z_n(\xi\ \mu)+C_nF_n(\xi, \mu, \sigma_2)=0, \quad \sigma_2\leqslant\xi\leqslant q, \tag{15.6}$$

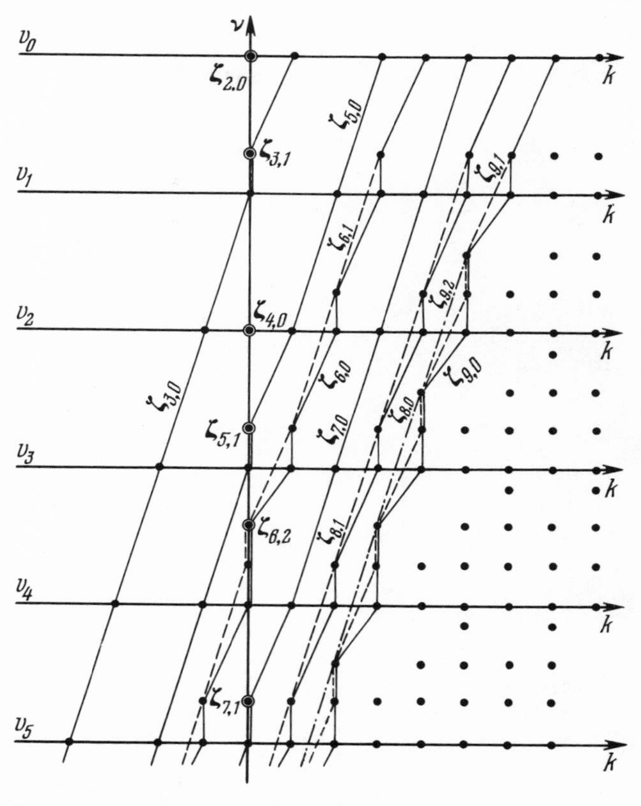

Fig. 40

and is negative at each point of the curve

$$\mathcal{K}_2(\xi, \eta) \equiv \eta - Z_n(\xi, \mu) - C_n F_n(\xi, \mu, \sigma_2) = 0, \quad \sigma_2 \leqslant \xi \leqslant q. \qquad (15.7)$$

We prove this under the assumption that $n \neq 3$; the change in the reasoning for $n = 3$ is trivial. For $n \neq 3$, the derivative in (14.2), at any point of the curve (15.6), is

$$\left.\frac{d\mathcal{K}_1}{d\xi}\right|_{(14.2)} = \mu^3 \frac{\tilde{\gamma}(\xi, Z_n(\xi, \mu) - C_n F_n(\xi, \mu, \sigma_2))}{\xi^2 + Z_n(\xi, \mu) - C_n F_n(\xi, \mu, \sigma_2)}$$

$$- \frac{dZ_n(\xi, \mu)}{d\xi} + C_n \mu^{n+1} \sum_{\nu=0}^{\pi(n-1)-1} \frac{1}{\xi^{n-\nu}} \ln^\nu \frac{1}{\mu}.$$

We shall find the dominant term of this expression for small ε, assuming that $\sigma_2 \leq \xi \leq q$, where $\sigma_2 = \varepsilon^{\lambda_2}$, $0 < \lambda_2 < 1/3$ [see (14.14)].

Using (15.1) and the inequalities

$$0 < F_n(\xi, \mu, \sigma_2) < 2\frac{\mu^{n+1}}{\sigma_2^{n-1}}, \quad \sigma_2 \leqslant \xi \leqslant q, \quad n \geqslant 2, \ n \neq 3,$$

which are obvious for sufficiently small ε, we obtain

$$\mu^3 \frac{\tilde{\gamma}(\xi, Z_n(\xi, \mu) - C_n F_n(\xi, \mu, \sigma_2))}{\xi^2 + Z_n(\xi, \mu) - C_n F_n(\xi, \mu, \sigma_2)} = \mu^3 \frac{\tilde{\gamma}(\xi, Z_n(\xi, \mu))}{\xi^2 + Z_n(\xi, \mu)} + O(\mu^{n+4-3\lambda_2(n+3)}); \tag{15.8}$$

the remainder term in this relation is uniformly bounded for $\sigma_2 \leq \xi \leq q$. It follows from the process of constructing the functions $\zeta_{n,\nu}(\xi)$ [see (14.6) and (14.7)] and (14.5) that

$$\frac{dZ_n(\xi, \mu)}{d\xi} = \mu^3 \frac{\tilde{\gamma}(\xi, Z_n(\xi, \mu))}{\xi^2 + Z_n(\xi, \mu)} - \frac{d}{d\xi}\mu^{n+1} w_{n+1} + O(\mu^{n+2-3\lambda_2(n+1)}). \tag{15.9}$$

To estimate the magnitude of the remainder term in (15.9), we use the inequality

$$\left|\frac{d}{d\xi}\mu^n w_n\right| \leqslant K\frac{\mu^n}{\sigma_2^{n-1}}, \qquad K = \text{const}, \quad \sigma_2 \leqslant \xi \leqslant q, \quad n = 2, 4, 5, \ldots.$$

This inequality, which holds for sufficiently small ε, is proved by using (15.2); it is easily seen that the remainder term is of strictly higher order than the second term on the right for $\varepsilon \to 0$ on the interval $\sigma_2 \leq \xi \leq q$.

Relations (15.8), (15.9), and the fact that $\zeta'_{n+1,\pi(n-1)}(\xi) \equiv 0$ [see (14.10)] imply that

$$\left.\frac{d\mathcal{K}_1}{d\xi}\right|_{(14.2)} = \mu^{n+1} \sum_{\nu=0}^{\pi(n-1)-1} \frac{1}{\xi^{n-\nu}} [\xi^{n-\nu}\zeta'_{n+1,\nu}(\xi) + C_n] \ln^\nu \frac{1}{\mu} + O(\mu^{n+2-3\lambda_2(n+1)}), \quad n \neq 3,$$

if $0 < \lambda_2 < 1/3$; the first term on the right is dominant for small ε. It follows from (15.2) that, if the constant $C_n > 0$ is large enough, this term is strictly positive for $\sigma_2 \leq \xi \leq q$; this proves the assertion of the lemma for the curve (15.6). The reasoning is similar in the case of the curve (15.7).

Lemma 9 implies the following result concerning the use of partial sums of the series (14.4) as asymptotic approximations for sections of trajectories of the system (5.6).

Theorem 5. Let λ_2 be any number such that $0 < \lambda_2 < 1/3$, let $Z_n(\xi,\eta)$, $n \geq 2$, be defined by (14.12), and let $\eta = \eta^*(\xi,\mu)$ be a solution of (14.2) with an initial value for $\xi = \sigma_2$, $\sigma_2 = \varepsilon^{\lambda_2}$ satisfying, for sufficiently small ε, either

$$|\eta^*(\sigma_2, \mu) - Z_n(\sigma_2, \mu)| < C\frac{\mu^{n+1}}{\sigma_2^{n-1}}, \quad n \neq 3, \quad C > 0, \qquad (15.10)$$

or (for $n = 3$)

$$|\eta^*(\sigma_2, \mu) - Z_3(\sigma_2, \mu)| < C\left(\mu^4 + \frac{\mu^5}{\sigma_2^3}\right), \quad C > 0.$$

Then, if ε is small enough, this solution is defined for $\sigma_2 \leq \xi \leq q$, $q = \text{const} > 0$,

$$\eta^*(\xi, \mu) = Z_n(\xi, \mu) + \Re_n(\xi, \varepsilon), \quad \sigma_2 \leqslant \xi \leqslant q,$$

here

$$|\Re_n(\xi, \varepsilon)| < \begin{cases} C_n \varepsilon^{\frac{n+1}{3} - \lambda_2(n-1)} & \text{if } n \neq 3, \\ C_3 \varepsilon^{\min\left(\frac{4}{3}, \frac{5}{3} - 3\lambda_2\right)} & \text{if } n = 3, \end{cases} \quad C_n = \text{const} > 0.$$

uniformly on the interval $\sigma_2 \leq \xi \leq q$.

16. Proof of Asymptotic Representations for the Junction Part

In Theorems 3-5, we gave asymptotic representations for parts of trajectories of the system (5.6) corresponding to various parts of the interval $-q \leq \xi \leq q$ of variation of ξ, under the assumption that the initial point of each of these parts satisfied a certain condition. We now prove that the results we obtained are mutually compatible, and that the asymptotic approximation for the junction part of $\mathfrak{T}_\varepsilon$ has the required order of accuracy ε^a, $a > 0$, for $\varepsilon \to 0$, uniformly on the whole part.

We first assume that $a > 4/3$.

We start by obtaining the asymptotic representation of the function (10.15) for $-\sigma_2 \leq \xi \leq 0$; this function describes $\mathfrak{T}_\varepsilon$ in the left part of a neighborhood of the junction point, small together with ε. It follows from (10.16) and (10.18) that we will have the desired accuracy uniformly on the interval $-\sigma_1 \leq \xi \leq 0$ if λ_1 and the positive integer N_2 are such that

$$0 < \lambda_1 < 1/3, \quad 1 + \lambda_1 N_2 \geqslant a. \tag{16.1}$$

To construct the asymptotic approximation of the function (6.5) describing the trajectory $\mathfrak{T}_\varepsilon$ at the initial part of the junction, we use (6.12) and (6.13). It is clear that we will have the required accuracy, uniformly for $-q \leq \xi \leq -\sigma_1$, if the positive integer N_1 is chosen so that

$$N_1 - \lambda_1(3N_1 - 2) \geqslant 1 + \lambda_1 N_2. \tag{16.2}$$

By virtue of Theorem 3, the asymptotic representation of the function (6.5) we are seeking exists if condition (7.9) is satisfied at the point $\xi = -q$, $q = \text{const}$, for $n = N_1 - 1$. However, it is clear from Sec. 7 that condition (7.9) is satisfied for the function (6.5) for all $n \geq 0$.

Theorem 4 implies that the desired asymptotic representation of the function (10.15) for $-\sigma_1 \leq \xi \leq 0$ exists if condition (11.18) is satisfied at the point $\xi = -\sigma_1$ for $n = N_2$. If, for the functions (6.5) and (10.15) describing successive parts of $\mathfrak{T}_\varepsilon$, the relation $\eta_{\mathfrak{T}}(-\sigma_1, \mu) = \tilde{\eta}_{\mathfrak{T}}(-\sigma_1, \varepsilon)$ is satisfied, then the quantity $\tilde{\eta}_{\mathfrak{T}}(-\sigma_1, \mu)$ needed for the verification of condition (11.18) can be calculated by using the asymptotic representation (6.12) of the function (6.5) obtained in the above analysis. Using (7.2) and (16.2), we obtain

$$\tilde{\eta}_{\mathfrak{T}}(-\sigma_1, \mu) = \eta_{\mathfrak{T}}(-\sigma_1, \varepsilon) = \sum_{m=0}^{N_1-1} \varepsilon^m \eta_m(-\sigma_1) + O(\varepsilon\sigma_1^{N_2}) =$$

$$= -\sigma_1^2 + \sum_{m=1}^{N_1-1} \sum_{k=0}^{\left]N_2+3m-2-\frac{m-1}{\lambda_1}\right]} \eta_k^m \frac{\varepsilon^m}{\sigma_1^{3m-k-2}} + O(\varepsilon\sigma_1^{N_2}). \tag{16.3}$$

It follows from (12.2) and (12.9) that the quantity $\mu_2 V_{N_2}(-\sigma_1/\mu, \mu)$ in condition (11.18) is given by the formula

$$\mu^2 V_{N_2}\left(-\frac{\sigma_1}{\mu}, \mu\right) = -\sigma_1^2 + \mu^2 z_0\left(-\frac{\sigma_1}{\mu}\right) + \sum_{k=1}^{N_2} \mu^{k+2} v_k\left(-\frac{\sigma_1}{\mu}\right) =$$

$$= -\sigma_1^2 + \sum_{k=0}^{N_2} \sum_{m=1}^{\left]\frac{1+\lambda_1 N_2 - \lambda_1 k - 2\lambda_1}{1-3\lambda_1}\right]} \gamma^m a_{m-1}^k \frac{\varepsilon^m}{\sigma_1^{3m-k-2}} + O(\varepsilon\sigma_1^{N_2}). \tag{16.4}$$

Changing the order of summation in (16.4), comparing the result with (16.3), and using (12.11), we see that condition (11.18) is satisfied.

We now obtain the asymptotic approximation of the function (14.3) describing $\mathfrak{T}_\varepsilon$ at the end of the junction part. It is evident from (14.13) and (14.14) that we will have the required accuracy, uniformly for $\sigma_2 \leq \xi \leq q$, if λ_2 and the positive number N_2 satisfy the inequalities

$$0<\lambda_2<\frac{1}{3}, \quad \frac{N_4+4}{3}-\lambda_2(N_4+2)\geqslant a. \tag{16.5}$$

Finally, we construct the asymptotic approximation of the function (10.15) for $0 \leq \xi \leq \sigma_2$; this function describes $\mathfrak{T}_\varepsilon$ in the right part of a neighborhood of the junction point that is small when ε is small. Relations (10.17) and (10.18) show that we will have the required accuracy uniformly for $0 \leq \xi \leq \sigma_2$ if the positive integer N_3 is such that

$$1+\lambda_2 N_3 \geqslant \frac{N_4+4}{3}-\lambda_2(N_4+2). \tag{16.6}$$

By virtue of Theorem 4, the required asymptotic representation of the function (10.15) exists on $0 \leq \xi \leq \sigma_2$ if N_2 and N_3 also satisfy the inequality

$$N_2 \geqslant N_3. \tag{16.7}$$

Theorem 5 implied that the asymptotic representation of the function (14.3) is possible if (15.10) is satisfied at the point $\xi = \sigma_2$ for $n = N_4 + 3$. Since the functions (10.15) and (14.3) describing successive parts of $\mathfrak{T}_\varepsilon$ satisfy $\eta^*_{\mathfrak{T}}(\sigma_2,\mu) = \tilde{\eta}_{\mathfrak{T}}(\sigma_2,\mu)$, the quantity $\eta^*_{\mathfrak{T}}(\sigma_2,\mu)$ needed for the verification of (15.10) can be calculated by using the asymptotic representation (10.17) (already justified) of the function (10.15). We conclude from (12.13) and (16.6) that

$$\eta^*_{\mathfrak{T}}(\sigma_2,\mu)=\tilde{\eta}_{\mathfrak{T}}(\sigma_2,\mu)=\sum_{m=0}^{N_3}\mu^{m+2}v_m\left(\frac{\sigma_2}{\mu}\right)+O\left(\frac{\mu^{N_4+4}}{\sigma_2^{N_4+2}}\right)=$$
$$=\sum_{m=0}^{N_3}\sum_{\nu=0}^{m}\sum_{k=3\nu+3}^{N_4+3}{}^{*}\, b^m_{k-m-2,\nu}\frac{\mu^k}{\sigma_2^{k-m-2}}\ln^\nu\frac{\sigma_2}{\mu}+\sum_{k=2}^{N_3+2} b^{k-2}_{0,\pi(k-2)}\mu^k\ln^{\pi(k-2)}\frac{\sigma_2}{\mu}+$$
$$+O\left(\frac{\mu^{N_4+4}}{\sigma_2^{N_4+2}}\right)=\sum_{k=2}^{N_4+3}\mu^k\left\{\sum_{\varkappa=0}^{\pi(k-2)-1}\left[\sum_{\nu=\varkappa}^{\pi(k-2)-1}\sum_{m=\nu}^{N_3}C^\varkappa_\nu b^m_{k-m-2,\nu}\frac{\ln^{\nu-\varkappa}\sigma_2}{\sigma_2^{k-m-2}}+\right.\right.$$
$$\left.\left.+C^\varkappa_{\pi(k-2)}b^{k-2}_{0,\pi(k-2)}\ln^{\pi(k-2)-\varkappa}\sigma_2\right]\ln^\varkappa\frac{1}{\mu}+b^{k-2}_{0,\pi(k-2)}\ln^{\pi(k-2)}\frac{1}{\mu}\right\}+O\left(\frac{\mu^{N_4+4}}{\sigma_2^{N_4+2}}\right). \tag{16.8}$$

By employing (15.3), we obtain the following formula for the quantity $Z_{N_4+3}(\sigma_2,\mu)$ occurring in condition (15.10):

$$Z_{N_4+3}(\sigma_2,\mu)=\sum_{k=2}^{N_4+3}\mu^k\sum_{\varkappa=0}^{\pi(k-2)}\zeta_{k,\varkappa}(\sigma_2)\ln^{\varkappa}\frac{1}{\mu}=$$
$$=\sum_{k=2}^{N_4+3}\mu^k\sum_{\varkappa=0}^{\pi(k-2)}\ln^{\varkappa}\frac{1}{\mu}\left[\sum_{\nu=0}^{\pi(k-2)-\varkappa-1}\sum_{m=2-k+\nu+\varkappa}^{N_3+2-k}\zeta_{m,\nu}^{k,\varkappa}\sigma_2^m\ln^{\nu}\sigma_2+\right.$$
$$\left.+\zeta_{0,\pi(k-2)-\varkappa}^{k,\varkappa}\ln^{\pi(k-2)-\varkappa}\sigma_2\right]+O\left(\frac{\mu^{N_4+4}}{\sigma_2^{N_4+2}}\right). \qquad (16.9)$$

By comparing (16.8) and (16.9) and using (15.5) and (14.11), we see that condition (15.10) is satisfied.

It is clear that, if $a > 4/3$, the numbers λ_1, λ_2, N_j, $j = 1,\ldots,4$, can be chosen to satisfy inequalities (16.1), (16.2), and (16.5)-(16.7), and we thus obtain an asymptotic expansion for the junction part (6.1) of $\mathfrak{T}_\varepsilon$ with given accuracy. For example, putting

$$\lambda_1=\lambda_2=1/7,\quad N_1=3,\quad N_2=N_3=7,\quad N_4=5,$$

we obtain a formula yielding an asymptotic approximation for the junction part with error of order less than ε^2. Moreover, with this choice of the numbers referred to above, there are many possibilities enabling us to select the most convenient formula for a given case.

By employing the lowest-order approximations indicated in (6.12), (10.16), (10.17), and (14.13), we can obtain asymptotic representations for the junction part (6.1), in which the number a characterizing the accuracy satisfies $a \leq 4/3$. The simplest is

$$\eta_{\mathfrak{T}}(\xi,\varepsilon)=\begin{cases}-\xi^2+O(\varepsilon^{2/3}), & -q\leqslant\xi\leqslant 0,\\ O(\varepsilon^{2/3}), & 0\leqslant\xi\leqslant q.\end{cases} \qquad (16.10)$$

Hence the curve (5.9) (a part of $\mathfrak{T}_0$) can serve as the zeroth approximation for the junction part of $\mathfrak{T}_\varepsilon$ under consideration. The following formula is more accurate [see (14.11) and (12.7)]:

$$\eta_{\mathfrak{T}}(\xi,\varepsilon)=\begin{cases}-\xi^2+\varepsilon\eta_1(\xi)+O(\varepsilon^a), & -q\leqslant\xi\leqslant-\varepsilon^{1/4},\\ -\xi^2+\mu^2 z_0\left(\frac{\xi}{\mu}\right)+O(\varepsilon^a), & -\varepsilon^{1/4}\leqslant\xi\leqslant 0,\\ \mu^2 v_0\left(\frac{\xi}{\mu}\right)+O(\varepsilon^a), & 0\leqslant\xi\leqslant\varepsilon^{1-a},\\ \Omega_0\mu^2+O(\varepsilon^a), & \varepsilon^{1-a}\leqslant\xi\leqslant q,\end{cases} \qquad (16.11)$$

where a is any number satisfying $2/3 < a < 1$. This formula shows that, in the immediate neighborhood of the junction point, trajectories $\mathfrak{T}_\varepsilon$ deviate from $\mathfrak{T}_0$ by an amount of lower order than ε. For example (16.11) for $\xi = 0$ yields

$$\eta_{\mathfrak{T}}(0, \varepsilon) = v_0(0)\gamma^{2/3}\varepsilon^{2/3} + O(\varepsilon^a), \quad 2/3 < a < 1.$$

We also have the following formula:

$$\eta_{\mathfrak{T}}(\xi, \varepsilon) = \begin{cases} \eta_0(\xi) + \varepsilon\eta_1(\xi) + O(\varepsilon^{4/3}), & -q \leqslant \xi \leqslant -\varepsilon^{1/9}, \\ \mu^2 v_0\left(\frac{\xi}{\mu}\right) + \mu^3 v_1\left(\frac{\xi}{\mu}\right) + \mu^4 v_2\left(\frac{\xi}{\mu}\right) + & \\ \quad + \mu^5 v_3\left(\frac{\xi}{\mu}\right) + O(\varepsilon^{4/3}), & -\varepsilon^{1/9} \leqslant \xi \leqslant \varepsilon^{1/9}, \\ \zeta_{2,0}(\xi)\mu^2 + \zeta_{3,1}(\xi)\mu^3 \ln\frac{1}{\mu} + & \\ \quad + \zeta_{3,0}(\xi)\mu^3 + O(\varepsilon^{4/3}), & \varepsilon^{1/9} \leqslant \xi \leqslant q; \end{cases} \quad (16.12)$$

this formula can be employed by using the expressions for the coefficients in it [see (5.9), (6.9), (12.17), and (14.11)]. Formula (16.12) can be proved without using the general relations (12.11) and (15.5); it is sufficient to refer to relations (6.9), (6.10), (12.1), (12.4), (12.8), (12.10), (12.16), and (14.11).

Theorem 6. If ε is sufficiently small, the junction part of $\mathfrak{T}_\varepsilon$ in the neighborhood of the junction point S has an asymptotic representation with a uniformly small error, of order lower than ε^a, where $a > 0$ is arbitrary. For $0 < a \leq 2/3$, it is sufficient to use the zeroth approximation (16.10); for $2/3 < a \leq 4/3$, formula (16.11) or formula (16.12) can be used; for $a > 4/3$, the required representation is described by relations (6.12), (6.13), (10.16)-(10.18), (14.13), and (14.14), where the numers λ_1, λ_2, N_1, N_2, N_3, and N_4 must be selected to satisfy (16.1), (16.2), and (16.5)-(16.7). The coordinate change (5.3) must be made. To find the coefficients in the asymptotic representation, integration of the nondegenerate system (1.1) is not necessary.

17. Asymptotic Approximations of the Trajectory on the Fast-Motion Part

In this section we investigate the fast-motion part SP of $\mathfrak{T}_0$; here $S(s_1,s_2)$ is the junction point and $P(p_1,p_2)$ is the subsequent drop point, on a stable part $x = \hat{x}_0(y)$ of the curve Γ (Fig. 41; the diagrams are similar for other types of junction point S shown in Figs. 26 and 27). Thus

$$p_2 = s_2, \quad s_1 \operatorname{sign} f''_x(S) < p_1 \operatorname{sign} f''_x(S).$$

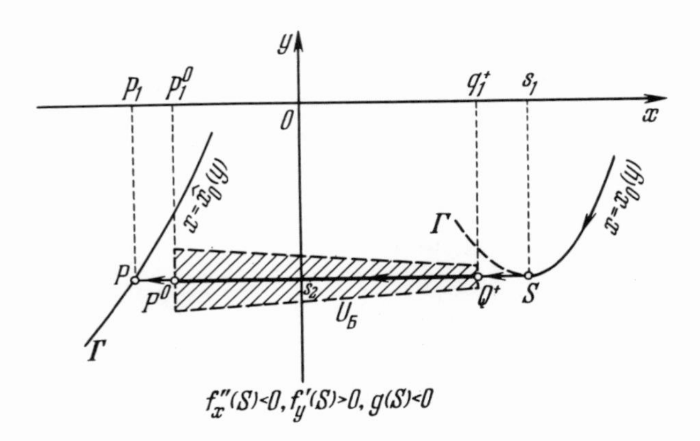

Fig. 41

On the horizontal segment SP, we consider a point $Q^+(q_1^+, q_2^+)$ as in Sec. 5, and a point $P^0(p_1^0, p_2^0)$ at a small finite distance from the drop point P. The equation of the part Q^+P^0 of $\mathfrak{T}_0$ can be written as

$$y = s_2, \quad q_1 \operatorname{sign} f_x''(S) \leqslant x \operatorname{sign} f_x''(S) \leqslant p_1^0 \operatorname{sign} f_x''(S). \tag{17.1}$$

Since the segment (17.1) does not cut Γ, there is a finite neighborhood U_B of this segment, determined by the inequalities

$$q_1^+ \operatorname{sign} f_x''(S) \leqslant x \operatorname{sign} f_x''(S) \leqslant p_1^0 \operatorname{sign} f_x''(S), |y - s_2| \leqslant \delta |x - s_1|$$

(Fig. 41) and lying completely in the attraction region of the stable part $x = \hat{x}_0(y)$, at each point of which

$$|f(x, y)| \geqslant k > 0. \tag{17.2}$$

Since, for sufficiently small ε, the trajectory $\mathfrak{T}_\varepsilon$ is close to the trajectory $\mathfrak{T}_0$ (Sec. 2 and Fig. 30), $\mathfrak{T}_\varepsilon$ enters U_B through a part $x = q_1 = s_1 + \rho_{\operatorname{sign} f_x''(S)}$ of the boundary, and leaves the region through a part $x = p_1^0$ of the boundary. The segment of $\mathfrak{T}_\varepsilon$, in U_B will be called *a fast-motion part.*

Let $D(q_1^+, d)$ be the point of intersection of $\mathfrak{T}_\varepsilon$ with the line $x = q_1^+$ on the boundary of U_B (see Fig. 35; cf. Fig. 41). Since this is the initial point of a fast-motion part and is simultaneously an end of the preceding junction part, it follows from (14.3), (8.2), and (5.3) that

$$d = s_2 + \eta_{\mathfrak{x}}^* (q, \gamma^{1/3}\varepsilon^{1/3}) \operatorname{sign} g(S).$$

Formally calculating the sum of the series (14.4) for $\xi = q$ and transforming back from μ to the original parameter ε, we obtain

$$d = s_2 + \\ + \operatorname{sign} g(S) \sum_{n=2}^{\infty} \gamma^{n/3} \varepsilon^{n/3} \sum_{\varkappa=0}^{\pi(n-2)} \zeta_{n,\varkappa}(q) \left(\frac{1}{3}\right)^{\varkappa} \left[\ln \frac{1}{\varepsilon} - \ln \gamma\right]^{\varkappa} = \\ = s_2 + \sum_{n=2}^{\infty} \varepsilon^{n/3} \sum_{\nu=0}^{\pi(n-2)} d_{n,\nu} \ln^{\nu} \frac{1}{3}; \qquad (17.3)$$

the coefficients here are given by the formula

$$d_{n,\nu} = \operatorname{sign} g(S) \sum_{\varkappa=\nu}^{\pi(n-2)} \frac{(-1)^{\varkappa-\nu} C_{\varkappa}^{\nu}}{3^{\varkappa}} \gamma^{n/3} \ln^{\varkappa-\nu} \gamma \zeta_{n,\varkappa}(q). \qquad (17.4)$$

By virtue of (17.2), the coordinate x is monotonic for increasing t on the fast-motion part of $\mathfrak{T}_{\varepsilon}$. Hence the equation of this part can be written as

$$y = y_{\mathfrak{T}}(x, \varepsilon), \qquad (17.5)$$
$$q_1^{+} \operatorname{sign} f_x''(S) \leqslant x \operatorname{sign} f_x''(S) \leqslant p_1^0 \operatorname{sign} f_x''(S);$$

here x is the independent variable. The function in (17.5) is clearly the solution of the equation

$$\frac{dy}{dx} = \varepsilon \frac{g(x, y)}{f(x, y)} \equiv \varepsilon \hbar(x, y), \qquad (17.6)$$

equivalent to (1.1), with a smooth right-hand side, satisfying the condition

$$y(q_1^{+}, \varepsilon) = d. \qquad (17.7)$$

We use (17.6) and (17.7) to seek an asymptotic expansion of the fast-motion part determined by (17.5).

Consider the formal series

$$y = \sum_{n=0}^{\infty}{}^{*} \varepsilon^{n/3} \sum_{\nu=0}^{\pi(n-2)} y_{n,\nu}(x) \ln^{\nu} \frac{1}{\varepsilon}, \qquad (17.8)$$

formally satisfying (17.6) for x between s_1 and p_1 and satisfying condition (17.7). Using (17.8) in Eq. (17.6) and formally expanding the right side of the equation in a series of terms $\varepsilon^{n/3} \ln^{\nu}(1/\varepsilon)$, we compare the coefficients of powers of ε and then the powers of $\ln(1/\varepsilon)$ (cf. Sec. 14), to obtain

$$\begin{aligned}
&y'_{0,0}(x)=0, \quad y'_{2,0}(x)=0,\\
&y'_{3,1}(x)=0, \quad y'_{3,0}(x)=\hbar(x, y_{0,0}(x)), \quad y'_{4,0}(x)=0,\\
&y'_{n,\pi(n-2)}(x)=0, \quad n\geqslant 5,\\
&y'_{n,\nu}(x)=\sum_{k=1}^{\left[\frac{n-3}{2}\right]}\frac{\hbar_y^{(k)}(x, y_{0,0}(x))}{k!}W^{\nu}_{n,k}(x), \qquad n\geqslant 5, \quad \nu=0,1,\ldots,\pi(n-2)-1,
\end{aligned} \tag{17.9}$$

where the $W^{\nu}_{n,k}(x)$ are functions introduced by means of the following relations [see (14.8)]:

$$\sum_{\substack{i_1+\ldots+i_k=\\ =n-3,\\ i_j\geqslant 2}}\left[\sum_{\nu_1=0}^{\pi(i_1-2)}y_{i_1,\nu_1}(x)\ln^{\nu_1}\frac{1}{\varepsilon}\cdot\ldots\cdot\sum_{\nu_k=0}^{\pi(i_k-2)}y_{i_k,\nu_k}(x)\ln^{\nu_k}\frac{1}{\varepsilon}\right]=$$

$$=\sum_{\nu=0}^{\pi(n-2)-1}W^{\nu}_{n,k}(x)\ln^{\nu}\frac{1}{\varepsilon}, \quad n\geqslant 5. \tag{17.10}$$

It is clear from (17.10) that, if $n\geq 5$, the functions $W^{\nu}_{n,k}(x)$ can be expressed in terms of only the coefficients $y_{m,\nu}(x)$, $m\leq n-3$; hence Eq. (17.9) is a differential equation with separable variables. Using the series (17.8) in (17.7), employing (17.3), and equating coefficients of the quantities $\varepsilon^{n/3}\ln^{\nu}(1/\varepsilon)$, we find that

$$\begin{aligned}
&y_{0,0}(q_1^+)=s_2,\\
&y_{n,\nu}(q_1^+)=d_{n,\nu}, \quad n\geqslant 2, \ \nu=0,1,\ldots,\pi(n-2).
\end{aligned} \tag{17.11}$$

Equation (17.9), together with the appropriate initial conditions (17.11), successively and uniquely determine the coefficients $y_{n,\nu}(x)$:

$$\begin{aligned}
&y_{0,0}(x)=s_2,\\
&y_{2,0}(x)=d_{2,0},\\
&y_{3,1}(x)=d_{3,1}, \quad y_{3,0}(x)=d_{3,0}+\int_{q_1^+}^{x}\hbar(z, s_2)\,dz,\\
&y_{4,0}(x)=d_{4,0},\\
&y_{n,\pi(n-2)}(x)=d_{n,\pi(n-2)}, \quad n\geqslant 5,\\
&y_{n,\nu}(x)=d_{n,\nu}+\int_{q_1^+}^{x}\sum_{k=1}^{\left[\frac{n-3}{2}\right]}\frac{\hbar_y^{(k)}(z, s_2)}{k!}W^{\nu}_{n,k}(z)\,dz,\\
&\qquad n\geqslant 5, \quad \nu=0,1,\ldots,\pi(n-2)-1.
\end{aligned} \tag{17.12}$$

It is plain from our formulas that the $y_{n,\nu}(x)$ are defined and smooth for x between s_1 and p_1. We stress that each of these functions can be effectively calculated in terms of values of the right sides of the system (1.1) and some of their derivatives on the segment SP of $\mathfrak{T}_0$. All these functions can also be used with x between q_1^+ and p_1^0.

In Sec. 18 we shall see that *the partial sum*

$$Y_n(x, \varepsilon) = \sum_{k=0}^{n}{}^{*} \varepsilon^{k/3} \sum_{\nu=0}^{\pi(k-2)} y_{k,\nu}(x) \ln^\nu \frac{1}{\varepsilon} \tag{17.13}$$

of the series (17.8) is an asymptotic approximation, for $\varepsilon \to 0$, *of the fast-motion part (17.5) of* $\mathfrak{T}_\varepsilon$:

$$y_{\mathfrak{T}}(x, \varepsilon) = \begin{cases} s_2 + O(\varepsilon^{2/3}), \\ Y_{N+1}(x, \varepsilon) + O\left(\varepsilon^{\frac{N+2}{3}} \ln^{\pi(N)} \frac{1}{\varepsilon}\right), \end{cases} \tag{17.14}$$

$$q_1^+ \operatorname{sign} f_x''(S) \leqslant x \operatorname{sign} f_x''(S) \leqslant p_1^0 \operatorname{sign} f_x''(S),$$

where N *is any positive integer.* This shows, in particular, that the part (17.1) of $\mathfrak{T}_0$ can serve as a zeroth approximation for the fast-motion part of $\mathfrak{T}_\varepsilon$ under consideration. By using (17.12), (17.4), and (14.11), we can obtain higher-order approximations:

$$y_{\mathfrak{T}}(x, \varepsilon) = s_2 + b_{0,0}^0 \gamma^{2/3} \operatorname{sign} g(S) \varepsilon^{2/3} + \frac{1}{3} b_{0,1}^1 \gamma \operatorname{sign} g(S) \varepsilon \ln \frac{1}{\varepsilon} +$$
$$+ \left(-\frac{1}{3} b_{0,1}^1 \gamma \ln \gamma + b_{0,0}^1 \gamma\right) \operatorname{sign} g(S) \varepsilon +$$
$$+ \left[\operatorname{sign} g(S) \oint_0^q \frac{\gamma(\theta, 0)}{\theta^2} d\theta + \int_{q_1^+}^{x} \frac{g(z, s_2)}{f(z, s_2)} dz \right] \varepsilon + O(\varepsilon^{4/3}),$$

$$q_1^+ \operatorname{sign} f_x''(S) \leqslant x \operatorname{sign} f_x''(S) \leqslant p_1^0 \operatorname{sign} f_x''(S). \tag{17.15}$$

Since $b_{0,0}^0 > 0$ [see (12.7) and (9.9)], we conclude from (17.15) that the fast-motion part (17.5) of $\mathfrak{T}_\varepsilon$ does not intersect the segment Q^+P^0 of $\mathfrak{T}_0$; it is always above this segment if $g(S) > 0$ and always below it if $g(S) < 0$ (cf. Sec. 5).

18. Derivation of Asymptotic Representations for the Fast-Motion Part

We shall establish the asymptotic formula (17.14) for the fast-motion part (17.5) of $\mathfrak{T}_\varepsilon$.

Lemma 10. If the points Q^+ and P^0 on the segment SP are fixed, then, for $n = 2, 3, \ldots$ and $\nu = 0, 1, \ldots, \pi(n-2)$, there is a constant $M_{n,\nu} > 0$ such that

$$|y_{n,\nu}(x)| \leqslant M_{n,\nu}, \quad |y'_{n,\nu}(x)| \leqslant M_{n,\nu}, \\ q_1^+ \operatorname{sign} f''_x(S) \leqslant x \operatorname{sign} f''_x(S) \leqslant p_1^0 \operatorname{sign} f''_x(S). \tag{18.1}$$

This is proved by induction, using (17.12), (17.9), (17.2), and the fact that $f(x,y)$ and $g(x,y)$ and their derivatives are bounded in U_B. All functions $y_{n,\pi(n-2)}(x)$, $n \geqslant 2$, are constant.

Lemma 11. Put

$$F_n(x, \varepsilon) = \int_{s_1}^{x} f(z, s_2)\, dz \varepsilon^{\frac{n+1}{3}} \ln^{\pi(n-1)} \frac{1}{\varepsilon}, \quad n \geqslant 1.$$

For sufficiently small ε, there is a constant $C_n > 0$ such that the derivative in (1.1) is positive at each point of the curve

$$\mathcal{K}_1^\circ(x, y) \equiv y - Y_n(x, \varepsilon) + C_n F_n(x, \varepsilon) = 0, \\ q_1^+ \operatorname{sign} f''_x(S) \leqslant x \operatorname{sign} f''_x(S) \leqslant p_1^0 \operatorname{sign} f''_x(S),$$

and negative at each point of the curve

$$\mathcal{K}_2^\circ(x, y) \equiv y - Y_n(x, \varepsilon) - C_n F_n(x, \varepsilon) = 0, \\ q_1^+ \operatorname{sign} f''_x(S) \leqslant x \operatorname{sign} f''_x(S) \leqslant p_1^0 \operatorname{sign} f''_x(S).$$

Suppose that $n \geq 4$; otherwise the reasoning is similar. Then at all points of the curve $\mathcal{K}_1(x,y) = 0$, the derivative, by virtue of (1.1), is

$$\left.\frac{d\mathcal{K}_1}{dt}\right|_{(1.1)} = \left.\frac{dy}{dt}\right|_{(1.1)} - \frac{dY_n(x, \varepsilon)}{dx} \cdot \left.\frac{dx}{dt}\right|_{(1.1)} + C_n \frac{dF_n(x, \varepsilon)}{dx} \cdot \left.\frac{dx}{dt}\right|_{(1.1)} =$$

$$= \frac{1}{\varepsilon} f(x, Y_n(x, \varepsilon) - C_n F_n(x, \varepsilon)) \Big\{ \varepsilon \hbar (x, Y_n(x, \varepsilon) -$$

$$- C_n F_n(x, \varepsilon)) - \frac{dY_n(x, \varepsilon)}{dx} + C_n f(x, s_2) \varepsilon^{\frac{n+1}{3}} \ln^{\pi(n-1)} \frac{1}{\varepsilon} \Big\}.$$

In the light of (18.1),

$$f(x, Y_n(x, \varepsilon) - C_n F_n(x, \varepsilon)) = f(x, s_2) + O(\varepsilon^{2/3}),$$

$$\hbar(x, Y_n(x, \varepsilon) - C_n F_n(x, \varepsilon)) = \hbar(x, Y_n(x, \varepsilon)) + O\left(\varepsilon^{\frac{n+1}{3}} \ln^{\pi(n-1)} \frac{1}{\varepsilon}\right),$$

and the procedure applied in constructing the $y_{n,\nu}(x)$ [see

(17.9)] yields

$$\frac{dY_n(x,\varepsilon)}{dx}=\varepsilon\hbar(x,Y_n(x,\varepsilon))+O\left(\varepsilon^{\frac{n+1}{3}}\ln^{\pi(n-1)-1}\frac{1}{\varepsilon}\right),\quad n\geqslant 4.$$

Hence

$$\left.\frac{d\mathscr{K}_1}{dt}\right|_{(1.1)}=C_n f^2(x,s_2)\,\varepsilon^{\frac{n-2}{3}}\ln^{\pi(n-1)}\frac{1}{\varepsilon}+O\left(\varepsilon^{\frac{n-2}{3}}\ln^{\pi(n-1)-1}\frac{1}{\varepsilon}\right),\qquad n\geqslant 4,$$

and it follows that the derivative is strictly positive for sufficiently small ε. We also note that the functions $F_n(x,\varepsilon)$, $n = 0,2,3,\ldots$, are strictly positive if x is between q_1^+ and p_1^0.

Let Q^+ and P^0 be the points considered in Sec. 17 on the fast-motion part SP of $\mathfrak{T}_0$. Lemma 11 implies the following result.

Theorem 7. Let $S(s_1,s_2)$ be a junction point of Γ, let the functions $Y_n(x,\varepsilon)$, $n \geq 1$, be given by (17.13), and let $y = y(x,\varepsilon)$ be a solution of Eq. (17.6) with initial value for $x=s_1+\rho_{\operatorname{sign} f''_x(S)}=q_1^+$, satisfying the condition

$$|y(q_1^+,\varepsilon)-Y_n(q_1^+,\varepsilon)|<C\varepsilon^{\frac{n+1}{3}}\ln^{\pi(n-1)}\frac{1}{\varepsilon},\quad C>0,\qquad (18.2)$$

for sufficiently small ε. Then, if ε is small enough, this solution is defined on the interval determined by the inequalities

$$q_1^+\operatorname{sign} f''_x(S)\leqslant x\operatorname{sign} f''_x(S)\leqslant p_1^0\operatorname{sign} f''_x(S)\qquad (18.3)$$

and has the representation

$$y(x,\varepsilon)=Y_n(x,\varepsilon)+\Re_n(x,\varepsilon),$$
$$q_1^+\operatorname{sign} f''_x(S)\leqslant x\operatorname{sign} f''_x(S)\leqslant p_1^0\operatorname{sign} f''_x(S);$$

moreover,

$$|\Re_n(x,\varepsilon)|<C_n\varepsilon^{\frac{n+1}{3}}\ln^{\pi(n-1)}\frac{1}{\varepsilon},\quad C_n=\text{const}>0,$$

uniformly on the interval (18.3).

The function (17.5) describing the fast-motion part of $\mathfrak{T}_\varepsilon$, satisfies condition (18.2) for $n \geq 1$; this follows from the choice of the initial conditions (17.11). Hence Theorem 7 can be used to find asymptotic approximations for the fast-motion part of $\mathfrak{X}_\varepsilon$. This completes the proof of (17.14).

Formula (17.14) clearly can be used to find the fast-motion part of $\mathfrak{X}_\varepsilon$ with any desired accuracy. However, the corresponding result is more easily stated if, instead of the functions (17.13), we consider partial sums of a series (17.8) of somewhat different form,

$$\tilde{Y}_n(x, \varepsilon) = \sum_{k=0}^{n-1}{}^{*} \varepsilon^{k/3} \sum_{\nu=0}^{\pi(k-2)} y_{k,\nu}(x) \ln^\nu \frac{1}{\varepsilon} + \varepsilon^{n/3} \sum_{\nu=1}^{\pi(n-2)} y_{n,\nu}(x) \ln^\nu \frac{1}{\varepsilon} .$$

We have the following *asymptotic representation of the fast-motion part (17.5) of* $\mathfrak{X}_\varepsilon$:

$$y_{\mathfrak{X}}(x, \varepsilon) = \tilde{Y}_{N+1}(x, \varepsilon) + O\left(\varepsilon^{\frac{N+1}{3}}\right),$$

$$q_1^+ \operatorname{sign} f''_x(S) \leqslant x \operatorname{sign} f''_x(S) \leqslant p_1^0 \operatorname{sign} f''_x(S);$$

here N is any positive integer. The proof uses reasoning similar to that used above.

Theorem 8. For sufficiently small ε, the fast-motion part (17.5) of $\mathfrak{X}_\varepsilon$ has the asymptotic representation

$$y_{\mathfrak{X}}(x, \varepsilon) = \tilde{Y}_{]3a+1]}(x, \varepsilon) + O(\varepsilon^a),$$

$$q_1^+ \operatorname{sign} f''_x(S) \leqslant x \operatorname{sign} f''_x(S) \leqslant p_1^0 \operatorname{sign} f''_x(S),$$

where $a \geq 2/3$ is arbitrary; for $0 < a < 2/3$ the zeroth approximation can be used. Calculation of the functions $\tilde{Y}_{]3a+1]}(x, \varepsilon)$ does not require integration of the nondegenerate system (1.1).

19. Special Variables for the Drop Part

Formulas (17.12) show that the functions $y_{n,\nu}(x)$, $n = 3, 5, 6, \ldots$, $\nu = 0, 1, \ldots, \pi(n-2) - 1$, increase unboundedly in absolute value when $x \to p_1$. Hence the series (17.8) cannot be considered as an asymptotic expansion for $\mathfrak{X}_\varepsilon$ on the whole interval of values of x between q_1^+ and p_1. To find asymptotic approximations for the part of $\mathfrak{X}_\varepsilon$, close to the drop point $P(p_1,p_2)$, we have to use special variables.

Suppose that the slow-motion part of $\mathfrak{X}_0$ following the fast-motion part SP is on a stable part $x = \hat{x}_0(y)$ of the curve Γ (Figs. 42 and 43; the situation for other types of junction point S is shown in Figs. 26 and 27). Let $\hat{S}(\hat{s}_1,\hat{s}_2)$

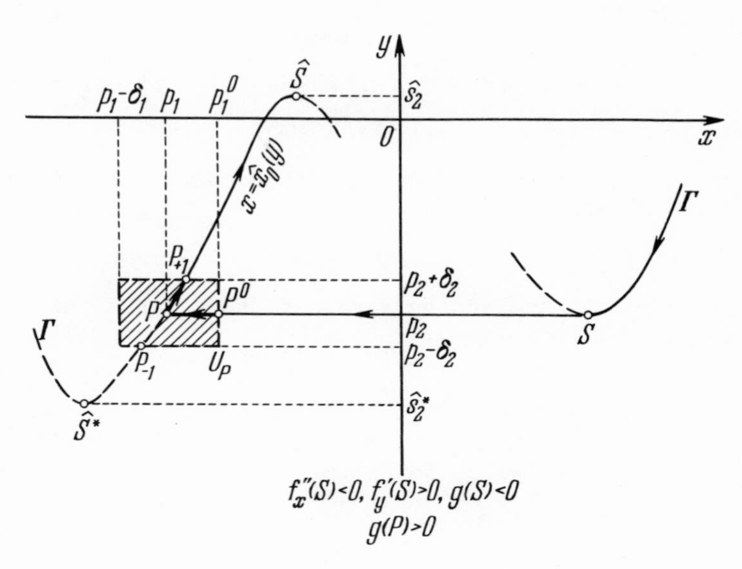

y
x
0
$\hat{S}$
$\hat{s}_2$
$p_1-\delta_1$
p_1
p_1^0
$x=\hat{x}_0(y)$
Γ
P_{+1}
P
P^0
P_{-1}
U_P
$p_2+\delta_2$
p_2
$p_2-\delta_2$
$\hat{s}_2^*$
$\hat{S}^*$
S
$f_x''(S)<0,\ f_y'(S)>0,\ g(S)<0$
$g(P)>0$

Fig. 42

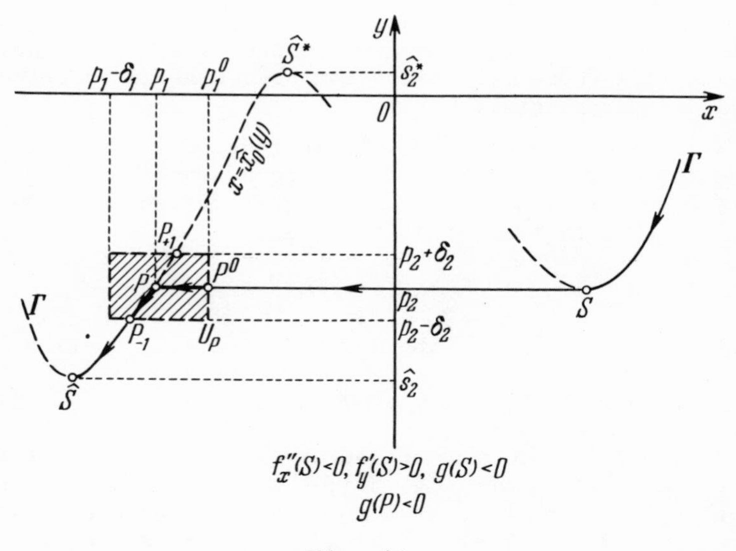

y
x
0
$\hat{S}^*$
$\hat{s}_2^*$
$p_1-\delta_1$
p_1
p_1^0
$x=\hat{x}_0(y)$
Γ
P_{+1}
P
P^0
P_{-1}
U_P
$p_2+\delta_2$
p_2
$p_2-\delta_2$
$\hat{s}_2$
$\hat{S}$
S
$f_x''(S)<0,\ f_y'(S)>0,\ g(S)<0$
$g(P)<0$

Fig. 43

be the junction point bounding the part $x = \hat{x}_0(y)$, and let $\hat{S}^*(\hat{s}_1^*, \hat{s}_2^*)$ be the second boundary of this part (cf. Sec. 3). Then

$$p_2 = s_2, \quad \hat{s}_2^* \operatorname{sign} g(P) < p_2 \operatorname{sign} g(P) < \hat{s}_2 \operatorname{sign} g(P),$$

$$\operatorname{sign} g(P) = \operatorname{sign} g(\hat{S}).$$

Let U_P be the small finite neighborhood of P defined by the inequalities

$$p_1 - \delta_1 \leqslant x \leqslant p_1 + \delta_1, \quad p_2 - \delta_2 \leqslant y \leqslant p_2 + \delta_2,$$

completely in the attraction region of the stable part $x = \hat{x}_0(y)$ and in which

$$|g(x, y)| \geqslant k > 0. \tag{19.1}$$

Furthermore, the stability of the part $x = \hat{x}_0(y)$ implies that $f'_x(\hat{x}_0(y), y) \leq -k < 0$ on any segment of this part; hence we assume that

$$\left[\frac{f'_x(x, y)}{g(x, y)} - \frac{f(x, y)\, g'_x(x, y)}{g^2(x, y)}\right] \operatorname{sign} g(P) \leqslant -k < 0 \tag{19.2}$$

in U_P. Finally, we assume that the curve $x = \hat{x}_0(y)$ cuts the horizontal parts $y = p_2 - \delta_2$ and $y = p_2 + \delta_2$ of the boundary of U_P at the points P_{-1} and P_{+1}, respectively (see Figs. 42 and 43). All these conditions can be satisfied by selecting δ_1 and δ_2 appropriately.

We take the point $P^0(p_1^0, p_2^0)$ introduced in Sec. 17 to be the intersection of SP with the boundary of the region U_P. The part $P^0 P P_{\operatorname{sign} g(P)}$ of $\mathfrak{T}_0$ in this region consists of the horizontal segment P^0P of the line

$$y = s_2, \quad p_1^0 \operatorname{sign} f''_x(S) \leqslant x \operatorname{sign} f''_x(S) \leqslant p_1 \operatorname{sign} f''_x(S), \tag{19.3}$$

and the arc $PP_{\operatorname{sign} g(P)}$ of the curve Γ with the equation

$$x = \hat{x}_0(y),\ p_2 \operatorname{sign} g(P) \leqslant y \operatorname{sign} g(P) \leqslant p_2 \operatorname{sign} g(P) + \delta_2. \tag{19.4}$$

If ε is small enough, the trajectory $\mathfrak{T}_\varepsilon$ is close to the trajectory $\mathfrak{T}_0$ (Sec. 2 and Fig. 30); hence $\mathfrak{T}_\varepsilon$ enters U_P through a part $x = p_1^0$ of the boundary, and leaves through the part $y = p_2 + \delta_2 \operatorname{sign} g(P)$ of the boundary. The segment of $\mathfrak{T}_\varepsilon$ in U_P will be called the *drop part*.

Let $E(p_1^0, e)$ be the point of intersection of $\mathfrak{T}_\varepsilon$ with the line $x = p_1^0$ lying on the boundary of U_P (Figs. 44 and 45; cf. Figs. 42 and 43). Since this is the initial point of a drop part and is simultaneously the end of the preceding fast-motion part, (17.5) implies that

$$e = y_{\mathfrak{T}}(p_1^0, \varepsilon).$$

By virtue of (19.1), the coordinate y on the drop part of $\mathfrak{T}_\varepsilon$ varies monotonically for increasing t; hence the equation of this part can be written as

$$x = \hat{x}_{\mathfrak{T}}(y, \varepsilon),$$
$$e \operatorname{sign} g(P) \leqslant y \operatorname{sign} g(P) \leqslant p_2 \operatorname{sign} g(P) + \delta_2; \tag{19.5}$$

the independent variable is y. The function (19.5) is clearly the solution of the following equation, equivalent to (1.1) [cf. (3.5)]:

$$\varepsilon \frac{dx}{dy} = \frac{f(x, y)}{g(x, y)} \equiv h(x, y); \tag{19.6}$$

here, the right-hand side is smooth and satisfies the conditions

$$x(e, \varepsilon) = p_1^0. \tag{19.7}$$

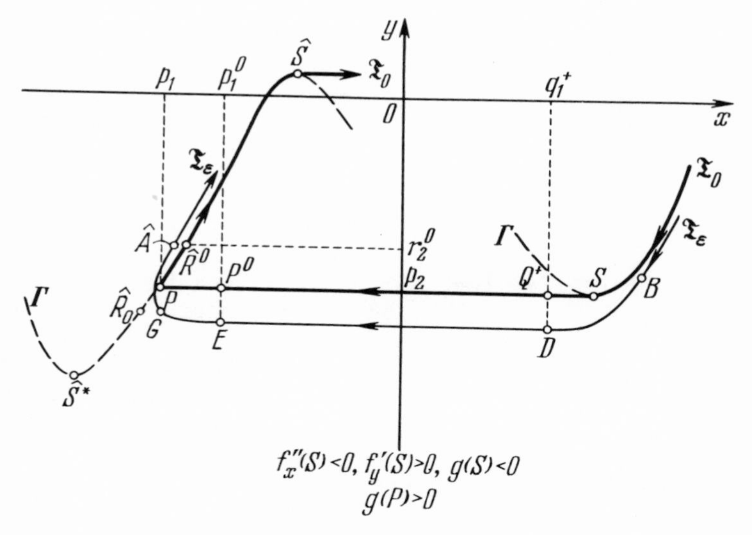

Fig. 44

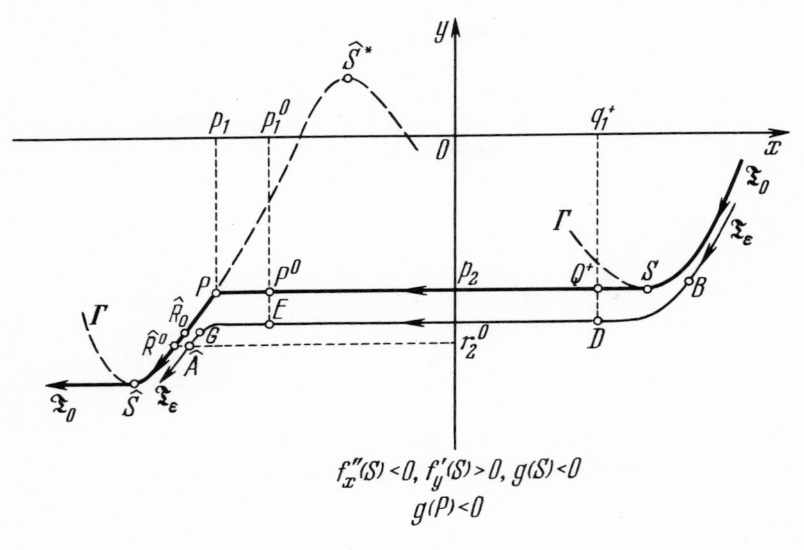

Fig. 45

Asymptotic approximations for the curve (19.5) are essentially different inside and outside a small neighborhood of the initial value $y = e$. Asymptotic expansions for the drop part of $\mathfrak{T}_\varepsilon$ outside such a neighborhood will be obtained by using (19.6), while in the immediate neighborhood of $y = e$ we must use a new change of variable.

We transform the original variables x, y, and t into new variables z, w, and θ by means of the formulas

$$x = p_1 + sz, \quad y = e + \varepsilon b \ln w, \quad t = \varepsilon\theta;$$
$$s = -\operatorname{sign} f_x''(S) = -\operatorname{sign} f(P^0), \quad b = -\frac{1}{h_x'(P)} = -\frac{g(P)}{f_x'(P)}; \tag{19.8}$$

here θ is a fast time (see Sec. 1), and the first two relations in (19.8) establish a one-to-one correspondence between the points of U_P in the (x,y) plane and the points of a region U_P^* of the (z,w) plane; the initial point E of the drop part corresponds to $z_0 = (p_1^0 - p_1)s > 0$, $w_0 = 1$. In the new variables, (1.1) becomes

$$\begin{cases} \dfrac{dz}{d\theta} = \varepsilon f(p_1 + sz,\ e + \varepsilon b \ln w), \\ \dfrac{dw}{d\theta} = \dfrac{w}{b} g(p_1 + sz,\ e + \varepsilon b \ln w); \end{cases} \tag{19.9}$$

this system does not have a small parameter multiplying a derivative.

By virtue of formulas (19.8), the drop part of $\mathfrak{T}_\varepsilon$ corresponds to the part of the trajectory of the system (19.9) in U_P^* and starting from the point (z_0, w_0). The second equation in (19.9) shows that, on this part, the coordinate w increases monotonically for increasing time θ. Hence the equation of this part is

$$z = z_{\mathfrak{T}}(w, \varepsilon), \quad 1 \leqslant w \leqslant \exp \frac{\delta_2 - (e - p_2) \operatorname{sign} g(P)}{\varepsilon |b|}; \qquad (19.10)$$

the independent variable is w. If we take w to be a parameter and return to the coordinates x and y, the curve (19.5), in parametric form, becomes

$$x = \boldsymbol{x}_{\mathfrak{T}}(w, \varepsilon) \equiv p_1 + s z_{\mathfrak{T}}(w, \varepsilon), \quad y = \boldsymbol{y}_{\mathfrak{T}}(w, \varepsilon) \equiv e + \varepsilon b \ln w, \qquad (19.11)$$

and w varies on the segment indicated in (19.10).

20. Asymptotic Approximations of the Drop Part of the Trajectory

We first find asymptotic representations for $\boldsymbol{x}_{\mathfrak{T}}(w, \varepsilon)$ and $\boldsymbol{y}_{\mathfrak{T}}(w, \varepsilon)$ as functions of the variable (the parameter) w. To this end, we must construct asymptotic approximations for $z_{\mathfrak{T}}(w, \varepsilon)$ on some interval of values of w.

The function (19.10) is clearly the solution of the following equation, which is equivalent to (19.9) [cf. (19.6)]:

$$\frac{dz}{dw} = \frac{sb}{w} h(p_1 + sz, \ e + \varepsilon b \ln w); \qquad (20.1)$$

here the right side is smooth and satisfies the condition

$$z(1, \varepsilon) = (p_1^0 - p_1) s \qquad (20.2)$$

[see (19.7)]. We use (20.1) and condition (20.2) to find asymptotic representations of the function (19.10), not on its whole domain, but only for $1 \leq w \leq 1/\varepsilon^a$, where $a > 0$. We also need an expansion for the formal value of the series (17.8) for $x = p_1^0$:

$$e = p_2 + \sum_{n=2}^{\infty} \varepsilon^{n/3} \sum_{\nu=0}^{\pi(n-2)} e_{n,\nu} \ln^\nu \frac{1}{\varepsilon}, \quad e_{n,\nu} = y_{n,\nu}(p_1^0). \qquad (20.3)$$

Consider the formal series

$$z = \sum_{n=0}^{\infty}{}^{*} \varepsilon^{n/3} \sum_{\nu=0}^{\pi(n-2)} z_{n,\nu}(w) \ln^\nu \frac{1}{\varepsilon}, \qquad (20.4)$$

formally satisfying (20.1) for $1 \leq w < \infty$ and condition (20.2). The coefficients in this series are determined as follows. We use (20.4) and (20.3) in (20.1), and formally expand the right side of the equation in a series of terms $\varepsilon^{n/3} \ln^{\nu} (1/\nu)$. It is convenient to carry out this expansion by using the formula

$$F\left(x_0 + \sum_{n=1}^{\infty} \lambda^n x_n,\ y_0 + \sum_{k=1}^{\infty} \lambda^k y_k\right) = F(x_0, y_0) +$$

$$+ \sum_{n=1}^{\infty} \lambda^n \sum_{k=1}^{n} \sum_{\substack{\alpha+\beta=k,\\ \alpha\geqslant 0,\ \beta\geqslant 0}} \frac{F_{x\ y}^{(\alpha)\,(\beta)}(x_0, y_0)}{\alpha!\beta!} \times \sum_{\substack{i_1+\ldots+i_\alpha+\\ +j_1+\ldots+j_\beta = n,\\ i_\nu \geqslant 1,\ j_\nu \geqslant 1}} x_{i_1} \ldots x_{i_\alpha} y_{j_1} \ldots y_{j_\beta}, \tag{20.5}$$

[cf. (3.7)], which can be verified directly. Using the notation

$$\begin{aligned} z_1 &= 0, \quad z_n(w) = \sum_{\nu=0}^{\pi(n-2)} z_{n,\nu}(w) \ln^{\nu} \frac{1}{\varepsilon}, \quad n \geqslant 2; \\ e_1 &= 0, \qquad e_n = \sum_{\nu=0}^{\pi(n-2)} e_{n,\nu} \ln^{\nu} \frac{1}{\varepsilon}, \qquad n \geqslant 2, \end{aligned} \tag{20.6}$$

and applying (20.5), we compare coefficients of powers of ε in Eq. (20.1) to obtain

$$z'_{0,0}(w) = \frac{sb}{w} h(p_1 + sz_{0,0}(w), p_2),$$

$$z'_n(w) = \sum_{m=0}^{\left]\frac{n-1}{2}\right]} \sum_{k=1}^{\left[\frac{n-3m}{2}\right]} \sum_{\substack{\alpha+\beta=k\\ \alpha\geqslant 0,\ \beta\geqslant 0}} \frac{h_{x\ y}^{(\alpha)\,(\beta+m)}(p_1 + sz_{0,0}(w), p_2)}{\alpha!\,\beta!\,m!} \times$$

$$\times \frac{s^{\alpha+1} b^{m+1} \ln^m w}{w} \sum_{\substack{i_1+\ldots+i_\alpha+\\ +j_1+\ldots+j_\beta = n-3m,\\ i_\nu \geqslant 2,\ j_\nu \geqslant 2}} z_{i_1} \ldots z_{i_\alpha} e_{j_1} \ldots e_{j_\beta} +$$

$$+ \begin{cases} 0, & \text{if } n \not\equiv 0 \pmod 3, \\ \dfrac{h_y^{(n/3)}(p_1 + sz_{0,0}(w), p_2)}{(n/3)!} \dfrac{sb^{\frac{n}{3}+1} \ln^{\frac{n}{3}} w}{w}, & \text{if } n \equiv 0 \pmod 3). \end{cases} \quad n \geqslant 2. \tag{20.7}$$

The function (12.14) satisfies the relations $\pi(n - 3m) = \pi(n) - m$ and

$$\max_{\substack{i_1+\ldots+i_k=m,\\ i_\nu\geqslant 2,\ m\geqslant 2}} \{\pi(i_1 - 2) + \ldots + \pi(i_k - 2)\} = \pi(m - 2) \tag{20.8}$$

[cf. (14.8)]; hence [see (20.6)]

$$\sum_{\substack{i_1+\ldots+i_\alpha+\\+j_1+\ldots+j_\beta=n-3m,\\ \alpha+\beta=k,\\ i_\nu\geqslant 2,\ j_\nu\geqslant 2,\ n-3m\geqslant 2}} z_{i_1}\ldots z_{i_\alpha} e_{j_1}\ldots e_{j_\beta} = \sum_{\nu=0}^{\pi(n-2)-m} W_{n,m,k}^{\alpha,\beta,\nu}(w)\ln^\nu\frac{1}{\varepsilon}. \quad (20.9)$$

Relation (20.9) shows that the right side of the nth equality in (20.7), for $n \geq 2$, is a polynomial in $\ln(1/\varepsilon)$ of degree $\pi(n-2)$ with coefficients depending on the functions $z_{k,\nu}(w)$, $k = 0,1,\ldots,n$. Here we compare coefficients of powers of $\ln(1/\varepsilon)$ in (20.7), and, using the series (20.4) in (20.2), we compare coefficients of identical terms $\varepsilon^{n/3}\ln^\nu(1/\varepsilon)$.

We note that $z_n(w)$ occurs only *once* in the right side of the nth relation in (20.7), in the term with $m = 0$, $k = 1$, $\alpha = 1$, $\beta = 0$. Hence, for $n \geq 2$, each of the functions $z_{n,\nu}(w)$, $\nu = 0,1,\ldots,\pi(n-2)$ occurs only *once* in the polynomial in $\ln(1/\varepsilon)$ referred to above, in the coefficient of $\ln^\nu(1/\varepsilon)$. Hence the procedure described previously yields

$$\begin{aligned} &z'_{0,0}(w)=\frac{sb}{w}h(p_1+sz_{0,0}(w),p_2), \quad z_{0,0}(1)=(p_1^0-p_1)s;\\ &z'_{2,0}(w)-z_{2,0}(w)\frac{bh'_x(p_1+sz_{0,0}(w),p_2)}{w}=\Phi_{2,0}(w)\equiv\\ &\equiv\frac{e_{2,0}sbh'_y(p_1+sz_{0,0}(w),p_2)}{w}, \quad z_{2,0}(1)=0; \qquad (20.10)\\ &z'_{n,\nu}(w)-z_{n,\nu}(w)\frac{bh'_x(p_1+sz_{0,0}(w),p_2)}{w}=\Phi_{n,\nu}(w),\\ &\qquad z_{n,\nu}(1)=0, \quad n\geqslant 2, \quad \nu=0,1,\ldots,\pi(n-2) \end{aligned}$$

(cf. Sec. 10). The right side $\Phi_{n,\nu}(w)$ in (20.10) for $n \geq 2$ can be expressed in terms of the functions $z_{0,0}(w)$, $z_{2,0}(w),\ldots$, $z_{n-2,\nu}(w)$ only, and does not contain the functions $z_{n-1,\kappa}(w)$, $z_{n,\kappa}(w)$. It is easily seen that each of the $\Phi_{n,\nu}(w)$, $n \geq 2$, depends on the constants $e_{2,0},\ldots,e_{n,y}$, but $e_{k,\kappa}$ with $k > n$ does not occur in these functions.

The first two relations in (20.10) are differential equations with separable variables and initial conditions. Solving this problem, we obtain the function

$$w=w(z)\equiv\exp\left\{\frac{f'_x(P)}{g(P)}\operatorname{sign} f''_x(S)\int_{(p_1^0-p_1)s}^{z}\frac{g(p_1+sx,p_2)}{f(p_1+sx,p_2)}dx\right\}. \qquad (20.11)$$

It can be verified directly that the function (20.11) is defined for $0 < z \leq (p_1^\circ - p_1)s$, takes values $w \geq 1$; it increases without bound when $z \to 0$, and its graph is monotonically decreasing and convex downward [see (19.2)]. The function

$$z = z_{0,0}(w) \tag{20.12}$$

[the initial coefficient in the series (20.4)] is the inverse of the function (20.11). The function (20.12) is clearly defined for $1 \leq w < \infty$, and it is monotonically decreasing and asymptotic to the w axis for $w \to \infty$ (Fig. 46).

The remaining pair of relations (20.10) forms initial-value problems for linear differential equations; hence we can successively and uniquely determine the coefficients $z_{n,\nu}(w)$ in the series (20.4):

$$\begin{gathered} z_{2,0}(w) = e_{2,0} sb\mathcal{N}(w) \int\limits_1^w \frac{h_y'(p_1 + sz_{0,0}(x), p_2)}{x\mathcal{N}(x)} dx, \\ z_{n,\nu}(w) = \mathcal{N}(w) \int\limits_1^w \frac{\Phi_{n,\nu}(x)}{\mathcal{N}(x)} dx, \qquad n \geqslant 2, \ \nu = 0, 1, \ldots, \pi(n-2); \end{gathered} \tag{20.13}$$

here we have used the notation

$$\mathcal{N}(w) = \exp \int\limits_1^w \frac{bh_x'(p_1 + sz_{0,0}(x), p_2)}{x} dx. \tag{20.14}$$

It is clear from the formulas we have obtained that the $z_{n,\nu}(w)$ are defined and smooth for $1 \leq w < \infty$. It should be stressed that each of these functions can be effectively calculated in terms of values of the right sides of (1.1) and certain of their derivatives on the segment $P^\circ P$ of $\mathfrak{T}_0$ and a finite number of initial coefficients in the expansion (20.3). This is important, because the exact value of e and its expansion (20.3) are not known, and we can use only

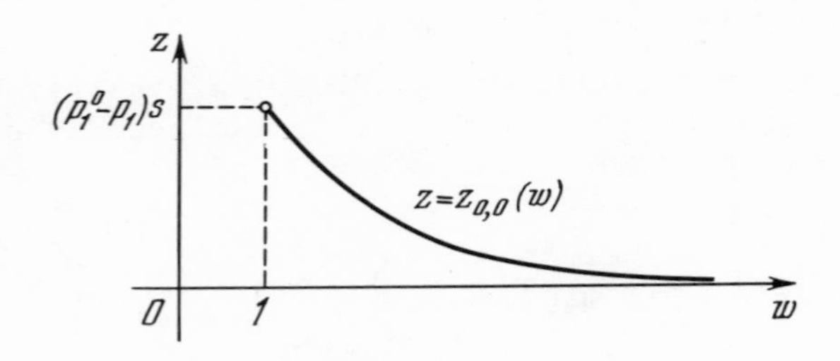

Fig. 46

partial sums

$$E_n(\varepsilon) = \sum_{k=0}^{n}{}^{*} \varepsilon^{k/3} \sum_{\nu=0}^{\pi(k-2)} e_{k,\nu} \ln^{\nu} \frac{1}{\varepsilon} \tag{20.15}$$

of the series (20.3) which serve as asymptotic approximations for this quantity.

We shall see in Sec. 21 that the partial sum

$$Z_n(w, \varepsilon) = \sum_{k=0}^{n}{}^{*} \varepsilon^{k/3} \sum_{\nu=0}^{\pi(k-2)} z_{k,\nu}(w) \ln^{\nu} \frac{1}{\varepsilon} \tag{20.16}$$

of the series (20.4) is an asymptotic approximation for $\varepsilon \to 0$ for the part of the solution (19.10) of problem (20.1), (20.2) on the interval defined by the inequalities $1 \le w \le 1/\varepsilon^{\alpha}$, where $\alpha > 0$ is arbitrary. Relations (19.11), on this interval of values of w, form the equation of $\mathfrak{T}_\varepsilon$ at the *beginning of the drop part*. We shall prove that *the initial drop part of* $\mathfrak{T}_\varepsilon$ *has the following asymptotic representations:*

$$\begin{cases} \boldsymbol{x}_{\mathfrak{T}}(w, \varepsilon) = p_1 + s z_{0,0}(w) + O(\varepsilon^{2/3}), \\ \boldsymbol{y}_{\mathfrak{T}}(w, \varepsilon) = p_2 + O(\varepsilon^{2/3}), \end{cases} \quad 1 \leqslant w \leqslant 1/\varepsilon^{2/3};$$

$$\begin{cases} \boldsymbol{x}_{\mathfrak{T}}(w, \varepsilon) = p_1 + s z_{0,0}(w) + s z_{2,0}(w)\, \varepsilon^{2/3} + O\left(\varepsilon \ln \frac{1}{\varepsilon}\right), \\ \qquad\qquad 1 \leqslant w \leqslant 1/\varepsilon; \\ \boldsymbol{y}_{\mathfrak{T}}(w, \varepsilon) = p_2 + e_{2,0}\varepsilon^{2/3} + O\left(\varepsilon \ln \frac{1}{\varepsilon}\right), \end{cases} \tag{20.17}$$

$$\begin{cases} \boldsymbol{x}_{\mathfrak{T}}(w, \varepsilon) = p_1 + s Z_{N+2}(w, \varepsilon) + O\left(\varepsilon^{\frac{N+3}{3}} \ln^{\pi(N+1)} \frac{1}{\varepsilon}\right), \\ \qquad\qquad 1 \leqslant w \leqslant 1/\varepsilon^{\frac{N+3}{3}}, \\ \boldsymbol{y}_{\mathfrak{T}}(w, \varepsilon) = E_{N+2}(\varepsilon) + \varepsilon b \ln w + O\left(\varepsilon^{\frac{N+3}{3}} \ln^{\pi(N+1)} \frac{1}{\varepsilon}\right), \end{cases}$$

where N is an arbitrary positive integer. In particular, we conclude that the part (19.3) of $\mathfrak{T}_0$ can serve as a zeroth approximation for $\mathfrak{T}_\varepsilon$ at the beginning of the drop part. The following is an example of a higher-order approximation:

$$\begin{cases} \boldsymbol{x}_{\mathfrak{T}}(w, \varepsilon) = p_1 + s z_{0,0}(w) + s z_{2,0}(w)\, \varepsilon^{2/3} + s z_{3,1}(w)\, \varepsilon \ln \frac{1}{\varepsilon} + s z_{3,0}(w)\, \varepsilon + O(\varepsilon^{4/3}), \\ \boldsymbol{y}_{\mathfrak{T}}(w, \varepsilon) = p_2 + e_{2,0}\varepsilon^{2/3} + e_{3,1}\varepsilon \ln \frac{1}{\varepsilon} + e_{3,0}\varepsilon + \varepsilon b \ln w + O(\varepsilon^{4/3}), \\ \qquad\qquad 1 \leqslant w \leqslant 1/\varepsilon^{4/3}. \end{cases} \tag{20.18}$$

On $\mathfrak{T}_\varepsilon$, consider the point $G(x,y)$ whose coordinates are given by (19.11) with the parameter values $w=\varepsilon^{-(N+3)/3}$. Let $\hat{R}_0$ be the point on the arc $\hat{S}^*\hat{S}$ of Γ with the same ordinate as G (Figs. 44 and 45). The point G is the end of the segment $\mathfrak{T}_\varepsilon$ of the trajectory EG at the beginning of the drop part. It remains to find the trajectory $\mathfrak{T}_\varepsilon$ at *the end of the drop part,* i.e., to find asymptotic representations of the function (19.5) on the segment

$$y \operatorname{sign} g(P) \leqslant y \operatorname{sign} g(P) \leqslant p_2 \operatorname{sign} g(P) + \delta_2.$$

To this end, we use Eq. (19.6) and the equation $x = \hat{x}_0(y)$ of the stable part $\hat{S}^*\hat{S}$ of the curve Γ.

Consider the formal power series

$$x = \hat{x}_0(y) + \varepsilon \hat{x}_1(y) + \ldots + \varepsilon^n \hat{x}_n(y) + \ldots \tag{20.19}$$

which *formally* satisfies Eq. (19.6) for y between $\hat{s}_2^*$ and s_2. The coefficients in this series are obtained as in Sec. 3; we need only take into account that we must start from the function $\hat{x}_0(y)$. We stress that each of the functions $\hat{x}_n(y)$ is defined and smooth for y between $\hat{s}_2^*$ and $\hat{s}_2$, and can be effectively calculated in terms of values of the right sides in (1.1) and certain of their derivatives on the part $\hat{S}^*\hat{S}$ of Γ.

In Sec. 21 we shall see that *the partial sum*

$$\hat{X}_n(y, \varepsilon) = \hat{x}_0(y) + \varepsilon \hat{x}_1(y) + \ldots + \varepsilon^n \hat{x}_n(y) \tag{20.20}$$

of the series (20.19) is an asymptotic approximation, for $\varepsilon \to 0$, *of the trajectory* $\mathfrak{T}_\varepsilon$ *at the end of the drop part:*

$$\hat{x}_{\mathfrak{T}}(y, \varepsilon) = \begin{cases} \hat{x}_0(y) + O(\varepsilon^{2/3}), \ p_2 \operatorname{sign} g(P) + O(\varepsilon^{2/3}) \leqslant y \operatorname{sign} g(P) \leqslant \\ \quad \leqslant p_2 \operatorname{sign} g(P) + \delta_2; \\ \hat{x}_0(y) + O\left(\varepsilon \ln \frac{1}{\varepsilon}\right), \ E_2(\varepsilon) \operatorname{sign} g(P) + \\ \quad + O\left(\varepsilon \ln \frac{1}{\varepsilon}\right) \leqslant y \operatorname{sign} g(P) \leqslant p_2 \operatorname{sign} g(P) + \delta_2; \\ \hat{X}_{\left]\frac{N+3}{3}\right[}(y, \varepsilon) + O\left(\varepsilon^{\frac{N+3}{3}} \ln^{\pi(N+1)} \frac{1}{\varepsilon}\right), \\ \left(E_{N+2}(\varepsilon) + \frac{N+3}{3} b\varepsilon \ln \frac{1}{\varepsilon}\right) \operatorname{sign} g(P) + O\left(\varepsilon^{\frac{N+3}{3}} \ln^{\pi(N+1)} \frac{1}{\varepsilon}\right) \leqslant \\ \quad \leqslant y \operatorname{sign} g(P) \leqslant p_2 \operatorname{sign} g(P) + \delta_2; \end{cases} \tag{20.21}$$

where N is an arbitrary positive integer. It follows, in

particular, that the part (19.4) of $\mathfrak{T}_0$ can serve as a zeroth approximation of $\mathfrak{T}_\varepsilon$ at the end of the drop part. A higher-order approximation is

$$\hat{x}_{\mathfrak{E}}(y, \varepsilon) = \hat{x}_0(y) + \varepsilon \frac{-f'_y(\hat{x}_0(y), y)\, g(\hat{x}_0(y), y)}{f'^2_x(\hat{x}_0(y), y)} + O(\varepsilon^{4/3}),$$

$$\left[p_2 + e_{2,0}\varepsilon^{2/3} + \left(e_{3,1} + \frac{4}{3}b\right)\varepsilon \ln\frac{1}{\varepsilon} + e_{3,0}\varepsilon\right] \operatorname{sign} g(P) + O(\varepsilon^{4/3}) \leqslant$$
$$\leqslant y \operatorname{sign} g(P) \leqslant p_2 \operatorname{sign} g(P) + \delta_2; \qquad (20.22)$$

cf. (3.14).

21. Proof of Asymptotic Representations for the Drop Part of the Trajectory

Lemma 12. The functions $z_{n,\nu}(w)$ in (20.12) and (20.13) have the asymptotic representations

$$z_{0,0}(w) = \frac{1}{w} \exp\left\{\frac{sf'_x(P)}{g(P)} \oint_0^{(p_1^0 - p_1)s} \frac{g(p_1 + sx, p_2)}{f(p_1 + sx, p_2)}\, dx\right\} + O\left(\frac{1}{w^2}\right) \qquad (21.1)$$
$$\text{for } w \to \infty;$$

$$z_{n,\nu}(w) = O(\ln^{\pi(n-2)-\nu} w) \quad \text{for } w \to \infty,$$
$$n \geqslant 2, \quad \nu = 0, 1, \ldots, \pi(n-2). \qquad (21.2)$$

Relation (21.1) is established by starting from the results in Sec. 13 and using the asymptotic properties of the function (20.11) for $z \to 0$:

$$w = \exp\left\{-\ln z + \frac{sf'_x(P)}{g(P)} \oint_0^{(p_1^0 - p_1)s} \frac{g(p_1 + sx, p_2)}{f(p_1 + sx, p_2)}\, dx + O(z)\right\}, \quad z \to 0.$$

Relations (21.2) are proved by induction. We first verify that the function $\mathcal{N}(w)$ given by (20.14) has the asymptotic representation

$$\mathcal{N}(w)^+ = \frac{1}{w} \exp \oint_1^\infty \frac{bh'_x(p_1 + sz_{0,0}(x), p_2)}{x}\, dx + O\left(\frac{1}{w^2}\right). \qquad (21.3)$$

The asymptotic behavior of $z_{2,0}(w)$ for $w \to \infty$ is inferred directly from the first equality in (20.13) and (21.3). If we assume that the assertion of the lemma holds for $z_{k,\nu}(w)$,

$k < n$, we conclude from (20.7) and (20.9) that

$$\Phi_{n,\nu}(w)^+ = O\left(\frac{1}{w}\ln^{\pi(n-2)-\nu} w\right). \tag{21.4}$$

The representation (21.2) follows from (20.13), (21.3), and (21.4).

More detailed calculations show that $z_{0,0}(w)$ and $z_{n,\nu}(w)$, for $n \geq 2$ and $\nu = 0, 1, \ldots, \pi(n-2)$, have the asymptotic representations

$$\begin{aligned} z_{0,0}(w)^+ &= \sum_{k=1}^{\infty} \frac{z_{k,0}^{0,0}}{w^k}; \\ z_{n,\nu}(w)^+ &= \sum_{\varkappa=0}^{\pi(n-2)-\nu} z_{0,\varkappa}^{n,\nu} \ln^{\varkappa} w + \sum_{k=1}^{\infty} \sum_{\varkappa=0}^{\pi(n-2)-\nu+1} \frac{z_{k,\varkappa}^{n,\nu} \ln^{\varkappa} w}{w^k}; \end{aligned} \tag{21.5}$$

there are formulas for the successive calculation of the coefficients $z_{n,\kappa}^{n,\nu}$, but they are very long and cumbersome. However, it is easily verified that

$$\begin{aligned} z_{2,0}(w)^+ &= e_{2,0} s b h_y'(P) + O\left(\frac{1}{w}\ln w\right), \\ z_{3,1}(w)^+ &= e_{3,1} s b h_y'(P) + O\left(\frac{1}{w}\ln w\right), \\ z_{3,0}(w)^+ &= s b^2 h_y'(P) \ln w + (e_{3,0} s b - s b^2) h_y'(P) + O\left(\frac{1}{w}\ln^2 w\right). \end{aligned} \tag{21.6}$$

There are results for the functions $\hat{x}_i(y)$ similar to Lemma 12. We also note that the functions $\hat{x}_i(y)$ and their derivatives for $y = p_2$ can be effectively calculated in terms of partial derivatives of $f(x,y)$ at the drop point P. This enables us to obtain formulas for the successive calculation of the coefficients $\hat{x}_{n,\nu}^{\kappa}$ in the expansion that is the formal value of the series (20.19) for $y = e + \varepsilon b \ln w$:

$$\sum_{i=0}^{\infty} \varepsilon^i \hat{x}_i(e + \varepsilon b \ln w) = p_1 + \sum_{n=2}^{\infty} \varepsilon^{n/3} \sum_{\nu=0}^{\pi(n-2)} \sum_{\varkappa=0}^{\pi(n-2)-\nu} \hat{x}_{n,\nu}^{\varkappa} \ln^{\varkappa} w \ln^{\nu} \frac{1}{\varepsilon}. \tag{21.7}$$

This expansion follows easily from (20.3), (3.7), and (20.8).

The following direct relation between the coefficients of the series (21.5) and (21.7) is important:

$$\begin{gathered} s z_{0,\varkappa}^{n,\nu} = \hat{x}_{n,\nu}^{\varkappa}, \\ n \geqslant 2, \quad \nu = 0, 1, \ldots, \pi(n-2), \quad \varkappa = 0, 1, \ldots, \pi(n-2) - \nu. \end{gathered} \tag{21.8}$$

The proof of (21.8) uses general formulas for the corresponding coefficients of the expansions (21.5) and (21.7).

Lemma 13. Let

$$F_n(w,\ \varepsilon)=\varepsilon^{\frac{n+1}{3}}\sum_{\nu=0}^{\pi(n-1)}(1+\ln^{\pi(n-1)-\nu}w)\ln^{\nu}\frac{1}{\varepsilon},\quad n\geqslant 1.$$

If ε is small enough, there is a constant $C_n > 0$ such that the derivative in (20.1) is positive at each point of the curve

$$\mathcal{K}_1(w,\ z)\equiv z-Z_n(w,\ \varepsilon)+C_nF_n(w,\ \varepsilon)=0,\qquad 1\leqslant w\leqslant 1/\varepsilon^a,\qquad (21.9)$$

and negative at each point of the curve

$$\mathcal{K}_2(w,\ z)\equiv z-Z_n(w,\ \varepsilon)-C_nF_n(w,\ \varepsilon)=0,\qquad 1\leqslant w\leqslant 1/\varepsilon^a;\qquad (21.10)$$

here $a > 0$ is arbitrary.

We assume that $n \geq 4$, and calculate the derivative in (20.1) at an arbitrary point of the curve $\mathcal{K}_1(w,z) = 0$:

$$\frac{d\mathcal{K}_1}{dw}\Big|_{(20.1)}=\frac{sb}{w}h(p_1+sZ_n(w,\ \varepsilon)-sC_nF_n(w,\ \varepsilon),\ e+\varepsilon b\ln w)-\\-\frac{dZ_n(w,\ \varepsilon)}{dw}+C_n\frac{dF_n(w,\ \varepsilon)}{dw}.$$

It follows from the process used in the construction of the functions $z_{n,\nu}(w)$ [see (20.7) and (20.10)] that

$$\frac{d\mathcal{K}_1}{dw}\Big|_{(20.1)}=\frac{1}{w}\varepsilon^{\frac{n+1}{3}}\Big\{[-bh'_x(p_1+sz_{0,0}(w),\ p_2)C_n+\\+w\Phi_{n+1,\pi(n-1)}(w)]\ln^{\pi(n-1)}\frac{1}{\varepsilon}+\\+\sum_{\nu=0}^{\pi(n-1)-1}[-bh'_x(p_1+sz_{0,0}(w),\ p_2)C_n(1+\ln^{\pi(n-1)-\nu}w)+\\+w\Phi_{n+1,\nu}(w)+(\pi(n-1)-\nu)C_n\ln^{\pi(n-1)-\nu-1}w]\ln^{\nu}\frac{1}{\varepsilon}\Big\}+\\+O\Big(\varepsilon^{\frac{n+2}{3}}\ln^{\pi(n)}\frac{1}{\varepsilon}\Big).\qquad (21.11)$$

By virtue of the properties of the function (20.12) and condition (19.2),

$$bh'_x(p_1+sz_{0,0}(w),\ p_2)\leqslant -k<0$$

for $w \geq 1$. Moreover, each of the functions $\Phi_{n+1,\nu}(w)$ is bounded on any finite interval of this half-line and, for

$w \to \infty$, has the asymptotic property indicated in (21.4). Hence all the expressions in square brackets in (21.11) are strictly positive for $1 \leq w \leq 1/\varepsilon^a$, with $a > 0$ if the constant $C_n > 0$ is large enough.

Since the initial point of the solution (19.10) of Eq. (20.1) is in the strip between the curves (21.9) and (21.10) [see (20.2) and (20.10)], Lemma 13 solves the problem of the feasibility of using partial sums of the series (20.4) as asymptotic approximations for the part of the solution corresponding to the interval $1 \leq w \leq 1/\varepsilon^a$, $a > 0$. We now return to the coordinates x, y, assuming that w is a parameter, and simultaneously using the asymptotic approximation (20.15) for e. This yields the following assertion concerning the asymptotic representation of trajectories of (1.1) at the beginning of a drop part.

Theorem 9. Suppose that $P(p_1,p_2)$ is an interior point of the stable part $\hat{S^*}\hat{S}$ of the curve Γ. The functions $Z_n(w,\varepsilon)$, $n \geq 1$, are given by (20.16), the $E_n(\varepsilon)$, $n \geq 1$, are defined by (20.15), and

$$x = \boldsymbol{x}(w, \varepsilon) \equiv p_1 + sz(w, \varepsilon), \qquad y = \boldsymbol{y}(w, \varepsilon) \equiv y^0 + \varepsilon b \ln w,$$

is the parametric solution of (19.6) with initial point for $w = 1$ satisfying the conditions

$$\begin{aligned} |z(1, \varepsilon) - (p_1^0 - p_1)s| &< C\varepsilon^{\frac{n+1}{3}} \ln^{\pi(n-1)} \frac{1}{\varepsilon}, \\ |y^0 - E_n(\varepsilon)| &< C\varepsilon^{\frac{n+1}{3}} \ln^{\pi(n-1)} \frac{1}{\varepsilon}, \end{aligned} \qquad C > 0, \qquad (21.12)$$

for sufficiently small ε. Then, this solution is defined for $1 \leqslant w \leqslant 1/\varepsilon^{n+1/3}$ and has the representation

$$\begin{aligned} \boldsymbol{x}(w, \varepsilon) &= p_1 + sZ_n(w, \varepsilon) + \Re_n^1(w, \varepsilon), \\ \boldsymbol{y}(w, \varepsilon) &= E_n(\varepsilon) + \varepsilon b \ln w + \Re_n^2(\varepsilon), \end{aligned} \qquad 1 \leqslant w \leqslant 1/\varepsilon^{\frac{n+1}{3}},$$

and

$$\begin{aligned} |\Re_n^1(w, \varepsilon)| &< C_n\varepsilon^{\frac{n+1}{3}} \ln^{\pi(n-1)} \frac{1}{\varepsilon}, \\ |\Re_n^2(\varepsilon)| &< C_n\varepsilon^{\frac{n+1}{3}} \ln^{\pi(n-1)} \frac{1}{\varepsilon}, \end{aligned} \qquad C_n = \text{const} > 0,$$

uniformly for $1 \leq w \leq 1/\varepsilon^{(n+1)/3}$.

In the light of the representation (17.14) and the choice of the initial conditions in (20.10), the function (19.11) describing $\mathfrak{T}_\varepsilon$ at the beginning of the drop part satisfies (21.12) for $n \geq 1$. Hence Theorem 9 can be used to obtain asymptotic representations of $\mathfrak{T}_\varepsilon$ at the initial drop part; this proves (20.17).

We shall also need a generalization of Lemma 2. Let $\hat{R}_0(\hat{r}_{01}, \hat{r}_{02})$ and $\hat{R}^0(\hat{r}_1^0, \hat{r}_2^0)$ be interior points of the arc $\hat{S}^*\hat{S}$, at a finite distance from both ends of this arc and from one another, and assume that $\hat{r}_{02}\operatorname{sign} g(\hat{S}) < \hat{r}_2^0 \operatorname{sign} g(\hat{S})$.

Lemma 14. If

$$F(y, \varepsilon) = \frac{\hat{r}_2^0 - y}{\hat{r}_2^0 - \hat{r}_{02}} \varepsilon^\alpha \ln^\beta \frac{1}{\varepsilon} + \frac{y - \hat{r}_{02}}{\hat{r}_2^0 - \hat{r}_{02}} \varepsilon, \qquad 0 < \alpha \leqslant 1, \quad \beta \geqslant 0, \tag{21.13}$$

then for sufficiently small ε there is a constant $C_n > 0$ such that the derivative in (1.1) is positive at all points of the curve

$$\mathcal{K}_1(y, x) \equiv x - \hat{X}_n(y, \varepsilon) + C_n \varepsilon^n F(y, \varepsilon) = 0,$$
$$\hat{r}_{02}\operatorname{sign} g(\hat{S}) \leqslant y \operatorname{sign} g(\hat{S}) \leqslant \hat{r}_2^0 \operatorname{sign} g(\hat{S}),$$

and negative at all points of the curve

$$\mathcal{K}_2(y, x) \equiv x - \hat{X}_n(y, \varepsilon) - C_n \varepsilon^n F(y, \varepsilon) = 0,$$
$$\hat{r}_{02}\operatorname{sign} g(\hat{S}) \leqslant y \operatorname{sign} g(\hat{S}) \leqslant \hat{r}_2^0 \operatorname{sign} g(\hat{S}).$$

The derivative in (1.1) can also be found at any point of the curve $\mathcal{K}_1(x, y) = 0$:

$$\left.\frac{d\mathcal{K}_1}{dt}\right|_{(1.1)} = \frac{1}{\varepsilon} g(\hat{X}_n(y, \varepsilon) - C_n \varepsilon^n F(y, \varepsilon), y) \Big\{ h(\hat{X}_n(y, \varepsilon) - C_n \varepsilon^n F(y, \varepsilon), y) - \varepsilon \frac{d\hat{X}_n(y, \varepsilon)}{dy} + C_n \varepsilon^{n+1} \frac{dF(y, \varepsilon)}{dy} \Big\}.$$

If $n \geq 1$ (for $n = 0$ the reasoning is similar),

$$g(\hat{X}_n(y, \varepsilon) - C_n \varepsilon^n F(y, \varepsilon), y) = g(\hat{x}_0(y), y) + O(\varepsilon),$$
$$h(\hat{X}_n(y, \varepsilon) - C_n \varepsilon^n F(y, \varepsilon), y) = h(\hat{X}_n(y, \varepsilon), y) - C_n \varepsilon^n F(y, \varepsilon) h'_x(\hat{x}_0(y), y) + O\left(\varepsilon^{n+1+\alpha} \ln^\beta \frac{1}{\varepsilon}\right)$$

(cf. Sec. 4). These equalities and a formula analogous to (4.5) show that

$$\left.\frac{d\mathcal{K}_1}{dt}\right|_{(1.1)} = -\varepsilon^{n-1} \{C_n F(y, \varepsilon) + \varepsilon \hat{x}_{n+1}(y)\} f'_x(\hat{x}_0(y), y) + O\left(\varepsilon^{n+\alpha} \ln^\beta \frac{1}{\varepsilon}\right).$$

Relations (19.2) and (21.13) show that the derivative is strictly positive for sufficiently small ε if the constant $C_n > 0$ is large enough.

Lemma 14 implies a result similar to Theorem 1, concerning the use of partial sums of the series (20.19) as asymptotic approximations for parts of trajectories of (1.1).

Theorem 10. Suppose that $x = \hat{x}_0(y)$ is a stable part of Γ, the functions $\hat{X}_n(y,\varepsilon)$, $n \geq 0$ are given by (20.20), and $x = \hat{x}(y,\varepsilon)$ is a solution of (19.6) with initial value for $y = \hat{r}_{02}$, satisfying

$$|\hat{x}(\hat{r}_{02},\varepsilon)-\hat{X}_n(\hat{r}_{02},\varepsilon)| < C\varepsilon^{n+\alpha}\ln^{\beta}\frac{1}{\varepsilon}, \quad C>0, \quad 0<\alpha\leqslant 1, \quad \beta\geqslant 0 \tag{21.14}$$

for sufficiently small ε. Then, if ε is small enough, this solution is defined on the interval determined by the inequalities

$$\hat{r}_{02}\operatorname{sign} g(\hat{S}) \leqslant y \operatorname{sign} g(\hat{S}) \leqslant \hat{r}_2^0 \operatorname{sign} g(\hat{S}) \tag{21.15}$$

and has the representation

$$\hat{x}(y,\varepsilon) = \hat{X}_n(y,\varepsilon) + \mathfrak{R}_n(y,\varepsilon),$$
$$\hat{r}_{02}\operatorname{sign} g(\hat{S}) \leqslant y \operatorname{sign} g(\hat{S}) \leqslant \hat{r}_2^0 \operatorname{sign} g(\hat{S}),$$

with

$$|\mathfrak{R}_n(y,\varepsilon)| < C_n\varepsilon^{n+\alpha}\ln^{\beta}\frac{1}{\varepsilon}, \quad C_n = \text{const} > 0, \tag{21.16}$$

uniformly on the interval (21.15).

We now take $\hat{R}_0$ to be the point defined in Sec. 20 and $\hat{R}^0$ to be the point $P_{\operatorname{sign} g(P)}$ (Sec. 19); then

$$\hat{r}_{02} = y, \qquad \hat{r}_2^0 = p_2 + \delta_2 \operatorname{sign} g(P)$$

(Figs. 42-45). We shall prove that the function (19.5) describing $\mathfrak{T}_\varepsilon$ at the end of the drop part satisfies (21.14). Fix an arbitrary positive integer N; we use the last two formulas in (20.17) [the reasoning is similar for the first two pairs of formulas (20.17)] to find the coordinates of the point G:

$$x = p_1 + sZ_{N+2}\left(1/\varepsilon^{\frac{N+3}{3}}, \varepsilon\right) + O\left(\varepsilon^{\frac{N+3}{3}}\ln^{\pi(N+1)}\frac{1}{\varepsilon}\right),$$
$$y = E_{N+2}(\varepsilon) + \frac{N+3}{3}b\varepsilon\ln\frac{1}{\varepsilon} + O\left(\varepsilon^{\frac{N+3}{3}}\ln^{\pi(N+1)}\frac{1}{\varepsilon}\right).$$

Hence, by virtue of the expansion (21.5),

$$\hat{x}_{\mathfrak{T}}(\boldsymbol{y}, \varepsilon) = \boldsymbol{x} = p_1 + s \sum_{n=0}^{N+2}{}^{*} \varepsilon^{n/3} \sum_{\nu=0}^{\pi(n-2)} z_{n,\nu}\left(\varepsilon^{-\frac{N+3}{3}}\right) \ln^{\nu} \frac{1}{\varepsilon} +$$

$$+ O\left(\varepsilon^{\frac{N+3}{3}} \ln^{\pi(N+1)} \frac{1}{\varepsilon}\right) = p_1 + \sum_{n=2}^{N+2} \varepsilon^{n/3} \times$$

$$\times \sum_{\nu=0}^{\pi(n-2)} \sum_{\varkappa=0}^{\nu} s z_{0,\varkappa}^{n,\nu-\varkappa} \left(\frac{N+3}{3}\right)^{\varkappa} \ln^{\nu} \frac{1}{\varepsilon} + O\left(\varepsilon^{\frac{N+3}{3}} \ln^{\pi(N+1)} \frac{1}{\varepsilon}\right).$$

But the expansion (21.7) implies that

$$\hat{X}_{\left]\frac{N+3}{3}\right]}(\boldsymbol{y}, \varepsilon) = \sum_{i=0}^{\left]\frac{N+3}{3}\right]} \varepsilon^{i} \hat{x}_{i}\left(E_{N+2}(\varepsilon) + \frac{N+3}{3} b\varepsilon \ln \frac{1}{\varepsilon}\right) +$$

$$+ O\left(\varepsilon^{\frac{N+3}{3}} \ln^{\pi(N+1)} \frac{1}{\varepsilon}\right) =$$

$$= p_1 + \sum_{n=2}^{N+2} \varepsilon^{n/3} \sum_{\nu=0}^{\pi(n-2)} \sum_{\varkappa=0}^{\nu} \hat{x}_{n,\nu-\varkappa}^{\varkappa} \left(\frac{N+3}{3}\right)^{\varkappa} \ln^{\nu} \frac{1}{\varepsilon} + O\left(\varepsilon^{\frac{N+3}{3}} \ln^{\pi(N+1)} \frac{1}{\varepsilon}\right).$$

It follows easily from relations (21.8) that condition (21.14) for the function (19.5) is satisfied for $n = [(N + 3)/3]$ at the corresponding α, β.

Hence Theorem 10 can be applied to find asymptotic approximations for the trajectory $\mathfrak{T}_\varepsilon$ at the end of the drop part; this proves relations (20.21). We note that, although $\boldsymbol{y}$ depends on ε, the distance between the points $\hat{R}_0$ and $P_{\operatorname{sign} g(P)}$ remains finite for small ε.

It is clear from (20.17) and (20.21) that it is possible to approximate the drop part of $\mathfrak{T}_\varepsilon$ with any desired accuracy with an error of order ε^{a}, $a > 0$. For example, if $a > 4/3$, we use the last pair of formulas (20.17), selecting N so that $N > 3a - 3$, and formula (20.21).

<u>Theorem 11.</u> For sufficiently small ε, the drop part of the trajectory $\mathfrak{T}_\varepsilon$, in the neighborhood of the drop point P, has an asymptotic representation with uniform accuracy up to small quantities of order lower than ε^{a}, where $a > 0$ is an arbitrary number. This representation is given by (20.17) and (20.21) with appropriate N, and calculation of the coefficients in the representation does not require integration of the nondegenerate system (1.1).

Let $\hat{A}$ be the point of intersection of $\mathfrak{T}_\varepsilon$ with the line $y = \hat{r}_2^0 = p_2 + \delta_2 \operatorname{sign} g(P)$, on the boundary of U_P (Figs. 44

and 45). A detailed analysis of the proofs of Theorem 10 and Lemma 14 shows that the remainder term satisfies, instead of (21.16), the inequality

$$|\mathfrak{R}_n(y, \varepsilon)| < C_n \varepsilon^n F(y, \varepsilon),$$

which holds on the interval (21.15). Hence (21.13) implies that

$$|\mathfrak{R}_n(\hat{r}_2^0, \varepsilon)| < C_n \varepsilon^{n+1},$$

and it follows that the abscissa of $\hat{A}$ satisfies a condition of the type (4.6). Since $\hat{A}$, which is the end of a drop part, is simultaneously the beginning of the succeeding slow-motion part, the results of Secs. 3 and 4 can be applied to this slow-motion part.

22. Asymptotic Approximations of the Trajectory for Initial Slow-Motion and Drop Parts

In our investigation of the fast-motion and drop parts in Secs. 17-21, we assumed that these parts follow a junction part. We shall briefly consider the asymptotic calculation of the initial fast-motion and drop parts of $\mathfrak{T}_\varepsilon$ (Sec. 2).

Suppose that the initial point $Q_0(x_0, y_0)$ of $\mathfrak{T}_\varepsilon$ is not on the curve Γ; we assume for simplicity that the coordinates of this point are independent of ε. Let $P_0(p_{01}, p_{02})$ be the drop point from Q_0 on a stable part $x = x_0(y)$ of the curve Γ [Fig. 47; the situation is similar for other positions of the

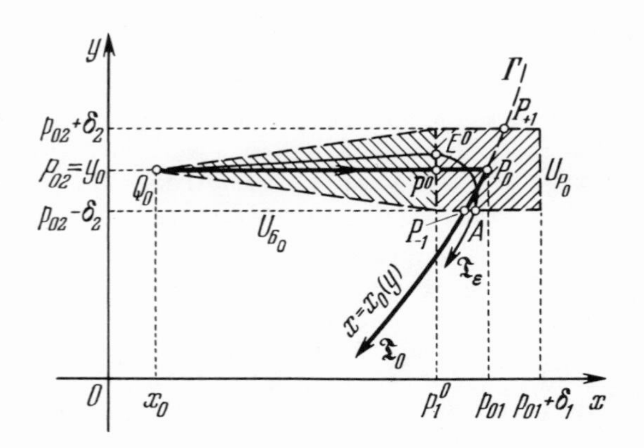

Fig. 47

vectors $f(Q_0)$ and $g(P_0)$]. On the horizontal segment Q_0P_0, we choose a point $P^\circ(p_1^\circ, p_2^\circ)$ at a small finite distance from the drop point P_0, and on the part $x = x_0(y)$ we choose points P_{+1} and P_{-1}, on opposite sides and at small finite distances from P_0.

It is clear that the part $Q_0P_0P_{\operatorname{sign} g(P_0)}$ of $\mathfrak{T}_0$ is formed of the horizontal segment Q_0P_0 with the equation

$$y = y_0, \quad x_0 \operatorname{sign} f(Q_0) \leqslant x \operatorname{sign} f(Q_0) \leqslant p_{01} \operatorname{sign} f(Q_0),$$

and the arc $P_0P_{\operatorname{sign} g(P_0)}$ of the curve Γ with the equation

$$x = x_0(y), \quad y_0 \operatorname{sign} g(P_0) \leqslant y \operatorname{sign} g(P_0) \leqslant y_0 \operatorname{sign} g(P_0) + \delta_2.$$

As in Secs. 17 and 19, we construct finite neighborhoods U_{B_0} and U_{P_0}, adjacent to one another with respect to a segment of the line $x = p_1^\circ$. The segment of $\mathfrak{T}_\varepsilon$, in U_{B_0} will be called the *initial fast-motion part*, and the segment of $\mathfrak{T}_\varepsilon$, in U_{P_0} will be called the *initial drop part*.

The equation of the initial fast-motion part is

$$\begin{gathered} y = y_{\mathfrak{T}}^0(x, \varepsilon), \\ x_0 \operatorname{sign} f(Q_0) \leqslant x \operatorname{sign} f(Q_0) \leqslant p_1^0 \operatorname{sign} f(Q_0); \end{gathered} \tag{22.1}$$

the function (22.1) is the solution of (17.6) satisfying

$$y(x_0, \varepsilon) = y_0. \tag{22.2}$$

We seek the asymptotic expansion of the function (22.1) as a formal power series

$$y = y_0(x) + \varepsilon y_1(x) + \ldots + \varepsilon^n y_n(x) + \ldots \tag{22.3}$$

[in contrast to the series (17.8)]. Using (22.3) in (17.6) and in (22.2) and carrying out the necessary calculations, we obtain the coefficients $y_i(x)$, $i = 0,1,2,\ldots$:

$$y_0(x) = y_0, \quad y_1(x) = \int_{x_0}^{x} \hbar(z, y_0)\, dz,$$

$$y_n(x) = \int_{x_0}^{x} \sum_{\nu=1}^{n-1} \frac{\hbar_y^{(\nu)}(z, y_0)}{\nu!} \sum_{\substack{i_1 + \ldots + i_\nu = n-1, \\ i_j \geqslant 1}} y_{i_1}(z) \ldots y_{i_\nu}(z)\, dz, \quad n \geqslant 2.$$

The following assertion is proved by reasoning similar to that used in Sec. 18. *The partial sum*

$$Y_n^0(x, \varepsilon) = y_0(x) + \varepsilon y_1(x) + \ldots + \varepsilon^n y_n(x)$$

of the series (22.3) is an asymptotic approximation, for $\varepsilon \to 0$, *of the initial fast-motion part (22.1) of* $\mathfrak{T}_\varepsilon$:

$$y_{\mathfrak{T}}^0(x, \varepsilon) = Y_{N-1}^0(x, \varepsilon) + O(\varepsilon^N),$$
$$x_0 \operatorname{sign} f(Q_0) \leqslant x \operatorname{sign} f(Q_0) \leqslant p_1^0 \operatorname{sign} f(Q_0);$$

where N is any positive integer.

An example is given by the formula

$$y_{\mathfrak{T}}^0(x, \varepsilon) = y_0 + \varepsilon \int_{x_0}^{x} \frac{g(z, y_0)}{f(z, y_0)} dz + O(\varepsilon^2),$$
$$x_0 \operatorname{sign} f(Q_0) \leqslant x \operatorname{sign} f(Q_0) \leqslant p_1^0 \operatorname{sign} f(Q_0).$$

We now consider the initial drop part. Let $E^\circ(p_1^\circ, e^\circ)$ be the point of intersection of $\mathfrak{T}_\varepsilon$ with the line $x = p_1^\circ$ (Fig. 47), so that $e^\circ = y_{\mathfrak{T}}^0(p_1^\circ, \varepsilon)$. We need the formal sum of the series (22.3) for $x = p_1^\circ$ and partial sums of this series

$$e^0 = y_0 + \sum_{n=1}^{\infty} e_n \varepsilon^n, \quad e_n = y_n(p_1^0); \quad E_n^0(\varepsilon) = y_0 + \sum_{k=1}^{n} e_k \varepsilon^k. \tag{22.4}$$

The equation of the initial drop part is

$$x = x_{\mathfrak{T}}^0(y, \varepsilon),$$
$$e^0 \operatorname{sign} g(P_0) \leqslant y \operatorname{sign} g(P_0) \leqslant y_0 \operatorname{sign} g(P_0) + \delta_2; \tag{22.5}$$

the function here is a solution of (19.6) and satisfies the condition $x(e^\circ, \varepsilon) = p_1^\circ$. Making the variable change

$$x = p_{01} + sz, \qquad y = e^0 + \varepsilon b \ln w,$$
$$s = -\operatorname{sign} f(P^0), \qquad b = -\frac{g(P_0)}{f'_x(P_0)},$$

[cf. (19.8)], we obtain the following equation and initial condition:

$$\frac{dz}{dw} = \frac{sb}{w} h(p_{01} + sz, \ e^0 + \varepsilon b \ln w), \quad z(1, \varepsilon) = (p_1^0 - p_{01}) s. \tag{22.6}$$

The solution $z = z_{\mathfrak{T}}^0(w, \varepsilon)$ of problem (22.6) corresponds to the following form of the curve (22.5):

$$x = \boldsymbol{x}^0_{\mathfrak{z}}(w, \varepsilon) \equiv p_{01} + sz^0_{\mathfrak{z}}(w, \varepsilon),$$
$$y = \boldsymbol{y}^0_{\mathfrak{z}}(w, \varepsilon) \equiv e^0 + \varepsilon b \ln w. \tag{22.7}$$

We seek the asymptotic expansion of $z = z^0_{\mathfrak{z}}(w, \varepsilon)$ as a formal power series:

$$z = z_0(w) + \varepsilon z_1(w) + \ldots + \varepsilon^n z_n(w) + \ldots \tag{22.8}$$

[in contrast to the series (20.4)]. Using (22.8) and (22.4) in (22.6) and employing (20.5), we find, after elementary calculations, that

$$z_0'(w) = \frac{sb}{w} h(p_{01} + sz_0(w), y_0), \quad z_0(1) = (p_1^0 - p_{01})s,$$
$$z_n'(w) - z_n(w)\frac{bh_x'(p_{01} + sz_0(w), y_0)}{w} = \Phi_n(w), \tag{22.9}$$
$$z_n(1) = 0, \quad n \geqslant 1,$$

where

$$\Phi_n(w) = -z_n(w)\frac{bh_x'(p_{01} + sz_0(w), y_0)}{w} + \frac{h_y^{(n)}(p_{01} + sz_0(w), y_0)}{n!}\frac{sb^{n+1}\ln^n w}{w} +$$
$$+ \sum_{m=0}^{n-1}\sum_{k=1}^{n-m}\sum_{\substack{\alpha+\beta=k,\\ \alpha\geqslant 0,\ \beta\geqslant 0}} \frac{h_x^{(\alpha)}{}_y^{(\beta+m)}(p_{01} + sz_0(w), y_0)}{\alpha!\beta!m!}\frac{s^{\alpha+1}b^{m+1}\ln^m w}{w} \times$$
$$\times \sum_{\substack{i_1+\ldots+i_\alpha+\\ +j_1+\ldots+j_\beta=n-m,\\ i_\nu\geqslant 1,\ j_\nu\geqslant 1}} z_{i_1}(w)\ldots z_{i_\alpha}(w)e_{j_1}\ldots e_{j_\beta}.$$

It is easily seen that the expression $\Phi_n(w)$ for $n \geq 1$ depends only on the functions $z_0(w)$, $z_1(w)$, ..., $z_{n-1}(w)$ (and only on the constants y_0, e_1, ..., e_n).

The first pair of relations in (22.9) determines the function

$$w = w(z) \equiv \exp\left\{\frac{f_x'(P_0)}{g(P_0)} \operatorname{sign} f(P^0) \int\limits_{(p_1^0 - p_{01})s}^{z} \frac{g(p_{01} + sx, y_0)}{f(p_{01} + sx, y_0)}dx\right\}$$

[cf. (20.11)], whose inverse is the initial coefficient $z_0(w)$ in the series (22.8). The other pairs of equations in (22.9) yield the coefficients $z_i(w)$, $i = 1, 2, \ldots$:

$$z_n(w) = \mathcal{N}(w)\int\limits_1^w \frac{\Phi_n(x)}{\mathcal{N}(x)}dx,$$

$$\mathcal{N}(w) = \exp\int\limits_1^w \frac{bh_x'(p_{01} + sz_0(x), y_0)}{x}dx, \quad n \geqslant 1.$$

The proof of the following assertion uses reasoning similar to that employed in Sec. 21. *If*

$$Z_n^0(w, \varepsilon) = z_0(w) + \varepsilon z_1(w) + \ldots + \varepsilon^n z_n(w)$$

is a partial sum of the series (22.8), we have the following asymptotic representation of the curve (22.7) on the intervals indicated:

$$\begin{cases} x_{\mathfrak{T}}^0(w, \varepsilon) = p_{01} + sz_0(w) + O\left(\varepsilon \ln \frac{1}{\varepsilon}\right), \\ y_{\mathfrak{T}}^0(w, \varepsilon) = y_0 + O\left(\varepsilon \ln \frac{1}{\varepsilon}\right), \end{cases} \quad 1 \leqslant w \leqslant 1/\varepsilon,$$

$$\begin{cases} x_{\mathfrak{T}}^0(w, \varepsilon) = p_{01} + sZ_N^0(w, \varepsilon) + O\left(\varepsilon^{N+1} \ln^{N+1} \frac{1}{\varepsilon}\right), \\ \qquad\qquad 1 \leqslant w \leqslant 1/\varepsilon^{N+1}, \\ y_{\mathfrak{T}}^0(w, \varepsilon) = E_N^0(\varepsilon) + \varepsilon b \ln w + O(\varepsilon^{N+1}); \end{cases}$$

here N is an arbitrary positive integer. An example is the representation

$$\begin{cases} x_{\mathfrak{T}}^0(w, \varepsilon) = p_{01} + sz_0(w) + sz_1(w)\varepsilon + O\left(\varepsilon^2 \ln^2 \frac{1}{\varepsilon}\right), \\ \qquad\qquad 1 \leqslant w \leqslant 1/\varepsilon^2. \\ y_{\mathfrak{T}}^0(w, \varepsilon) = y_0 + \varepsilon\left[\int\limits_{x_0}^{p_1^0} \frac{g(x, y_0)}{f(x, y_0)} dx - \frac{g(P_0)}{f'_x(P_0)} \ln w\right] + O(\varepsilon^2), \end{cases}$$

This formula and the corresponding results in Secs. 17 and 20 explain one essential difference between the behavior of $\mathfrak{T}_\varepsilon$ on fast-motion and drop parts following a junction part, on the one hand, and on initial fast-motion and drop parts on the other. In the situation in which

$$\operatorname{sign}[g(S)g(P)] = -1, \quad \operatorname{sign}[g(S)f(P^0)f'_y(P)] = +1$$

(Fig. 44 shows one of these cases), the fast-motion and drop parts of $\mathfrak{T}_\varepsilon$ do not intersect $\mathfrak{T}_0$. If

$$\operatorname{sign}[g(Q_0)g(P_0)] = -1, \quad \operatorname{sign}[g(Q_0)f(P^0)f'_y(P_0)] = +1$$

(Fig. 47 shows one of these cases), $\mathfrak{T}_\varepsilon$, on initial fast-motion and drop parts, intersects $\mathfrak{T}_0$ twice.

Now consider the formal power series (3.6) [cf (20.19)] for Eq. (19.6) [see (3.5)]. The calculation of the coefficients in this series is described in Sec. 3; the initial coefficient is taken to be the function in the equation $x = x_0(y)$ of the stable part of Γ (Fig. 47).

The reasoning in the proof of the following assertion is similar to the reasoning used in Sec. 21. *If (3.12) is a partial sum of the series (3.6), we have the following asymptotic representations of the function (22.5) on the intervals indicated:*

$$x_{\mathfrak{T}}^{0}(y,\varepsilon)=\begin{cases} x_0(y)+O\left(\varepsilon\ln\frac{1}{\varepsilon}\right), \\ y_0 \operatorname{sign} g(P_0)+O\left(\varepsilon\ln\frac{1}{\varepsilon}\right)\leqslant y\operatorname{sign} g(P_0)\leqslant y_0\operatorname{sign} g(P_0)+\delta_2, \\ X_N(y,\varepsilon)+O\left(\varepsilon^{N+1}\ln^{N+1}\frac{1}{\varepsilon}\right), \\ \left(E_N^0(\varepsilon)+(N+1)b\varepsilon\ln\frac{1}{\varepsilon}\right)\operatorname{sign} g(P_0)+ \\ +O(\varepsilon^{N+1})\leqslant y\operatorname{sign} g(P_0)\leqslant y_0\operatorname{sign} g(P_0)+\delta_2; \end{cases}$$

here N is any positive integer. The following is an example:

$$x_{\mathfrak{T}}^{0}(y,\varepsilon)=x_0(y)+\varepsilon\frac{-f_y'(x_0(y),y)\,g(x_0(y),y)}{f_x'^2(x_0(y),y)}+O\left(\varepsilon^2\ln^2\frac{1}{\varepsilon}\right),$$

$$\left[y_0-\frac{2g(P_0)}{f_x'(P_0)}\varepsilon\ln\frac{1}{\varepsilon}+\varepsilon\int_{x_0}^{p_1^0}\frac{g(x,y_0)}{f(x,y_0)}dx\right]\operatorname{sign} g(P_0)+O(\varepsilon^2)\leqslant$$
$$\leqslant y\operatorname{sign} g(P_0)\leqslant y_0\operatorname{sign} g(P_0)+\delta_2.$$

Denote the point $P_{\operatorname{sign} g(P_0)}$ by $R^{\circ}(r_1^{\circ},r_2^{\circ})$, and let A be the point of intersection of $\mathfrak{T}_{\varepsilon}$ with the line $y = r_2^{\circ} = \delta_2 \times \operatorname{sign} g(P_0)$ (Fig. 47). It is an important fact that the abscissa of A satisfies (4.6). Thus, to find the asymptotic representation of the slow-motion part of $\mathfrak{T}_{\varepsilon}$, following the initial drop part, we can apply results of Secs. 3 and 4.

CHAPTER III

SECOND-ORDER SYSTEMS. ALMOST-DISCONTINUOUS PERIODIC SOLUTIONS

A degenerate system can have closed phase trajectories corresponding to discontinuous periodic solutions. For sufficiently small values of the parameter, there is a stable limit cycle of the nondegenerate system close to each such trajectory of the degenerate system; the corresponding periodic solution is of relaxation-oscillation nature.

In this chapter we obtain asymptotic representations, with any given degree of accuracy, for the trajectory of an almost-discontinuous periodic solution of a second-order system with a small parameter multiplying a derivative. We also find asymptotic representations for the period of relaxation oscillations.

1. Existence and Uniqueness of an Almost-Discontinuous Periodic Solution

We consider phase trajectories of the nondegenerate system

$$\begin{cases} \varepsilon\dot{x} = f(x, y), \\ \dot{y} = g(x, y), \end{cases} \tag{1.1}$$

where x and y are scalar functions of the independent variable t, and ε is a small parameter; all assumptions made in Chapter II, Sec. 1 are retained.* Simple examples (Chapter I, Sec. 4) show that

$$\begin{cases} f(x, y) = 0, \\ \dot{y} = g(x, y) \end{cases} \tag{1.2}$$

*We give the chapter number in references to the preceding chapters.

can have *closed trajectories*. We note that the presence (or absence) of closed trajectories of (1.2) is established directly from the functions $f(x,y)$ and $g(x,y)$ without the necessity of solving (1.2) (cf. Chapter II, Sec. 2).

We assume that (1.2) has a closed phase trajectory Z_0. It follows from considerations in Chapter II, Sec. 2 that Z_0 is a continuous closed curve in the (x,y) plane formed of a finite number of alternating slow- and fast-motion parts (Fig. 48).

The solution $x = x(t)$, $y = y(t)$ of (1.2) corresponding to the curve Z_0 is a *discontinuous periodic solution* (Chapter I, Sec. 6). For increasing time t, the coordinate y of the solution varies continuously, but the coordinate x has discontinuities (of the first kind) for all values of t for which the phase point of (1.2) is a junction point of Z_0. The parts of the solution between adjacent discontinuity points are obtained as follows.

Suppose that, for $t = t^*$, the phase point of (1.2) is at a drop point $\tilde{P}(\tilde{p}_1,\tilde{p}_2)$ on a stable part $x = x_0(y)$ of the curve Γ (Fig. 48). Substitution of $x = x_0(y)$ in the second

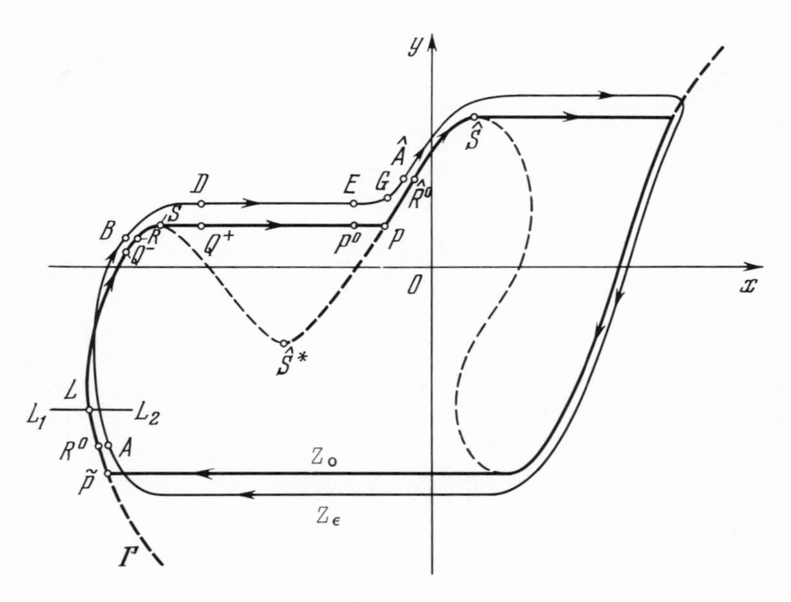

Fig. 48

of Eqs. (1.2) yields

$$y = g(x_0(y), y). \tag{1.3}$$

The condition $y(t^*) = \tilde{p}_0$ uniquely distinguishes a solution $y = y^*(t)$ of (1.3) for $t^* \leq t \leq t^{**}$, where t^{**} is the next discontinuity value following t^*, i.e., the time when the phase point of (1.2) reaches the junction point $S(s_1, s_2)$ which is the boundary for the part $x = x_0(y)$. The equation of this part can be used to find the function $x = x^*(t) \equiv x_0(y^*(t))$, $t^* \leq t \leq t^{**}$.

Since the fast-motion parts of Z_0 are traversed instantaneously, the period T_0 of the corresponding discontinuous periodic solution of the degenerate system (1.2) is equal to the sum of the traversal times for (1.2) of all slow-motion parts of the trajectory Z_0. The traversal time for the part $\widetilde{PS}$ is easily obtained from (1.3) (Fig. 48):

$$T_{\widetilde{PS}} = \int_{\tilde{p}_2}^{s_2} \frac{dy}{g(x_0(y), y)} = \int_{\widetilde{PS}} \frac{dy}{g(x, y)}; \tag{1.4}$$

it can be calculated similarly for the other slow-motion parts. The sum of these times is

$$T_0 = \oint_{Z_0} \frac{dy}{g(x, y)}, \tag{1.5}$$

where the integral is taken in the direction of motion on Z_0. It should be noted that the period T_0 of the discontinuous periodic solution can be effectively calculated in terms of the values of $g(x,y)$ on Z_0.

It is easily seen that, *for sufficiently small* ε, *at least one closed trajectory of the nondegenerate system (1.1) is close to* Z_0. To prove this, it is sufficient to construct an annular neighborhood of the closed curve Z_0 of small but *finite* width, such that the phase-velocity vectors of (1.1), for sufficiently small ε, are directed towards the interior of the neighborhood at all points of its boundary (cf. Chapter II, Sec. 2 and Fig. 30). However, we have the following stronger result [3, 19].

Theorem 1. Let the conditions of Chapter II, Sec. 1 be satisfied, and let the degenerate system (1.2) have a closed phase trajectory Z_0. Then, for sufficiently small ε, the nondegenerate system (1.1) has a unique and stable limit cycle Z_ε having as a zeroth approximation the curve Z_0, i.e., $Z_\varepsilon \to Z_0$ when $\varepsilon \to 0$.

In fact, consider the slow-motion part $\tilde{P}S$ of the closed trajectory Z_0 (Fig. 48) and, on it, consider a point L at a finite distance from both the drop point $\tilde{P}$ and the junction point S. Let L_1L_2 be the segment, parallel to the Ox axis, with midpoint at L, and with length $3M_1\varepsilon$, where M_1 is the constant in inequality (II.4.1) for $n = 1$. By using the asymptotic approximations constructed in Chapter II, it can easily be verified that a trajectory of (1.1) starting at any point of L_1L_2 again reaches L_1L_2 after a time differing only slightly from T_0 [see (1.5)]. Hence passage on trajectories of (1.1) generates a continuous mapping of L_1L_2 on itself. Since such a mapping always has a fixed point, the nondegenerate system (1.1) has at least one closed trajectory with the zeroth asymptotic approximation Z_0.

Let Z_ε be any one of the closed trajectories of (1.1) such that $Z_\varepsilon \to Z_0$ when $\varepsilon \to 0$. It can be proved that, if ε is small enough,

$$\oint\limits_{Z_\varepsilon} \left[\frac{1}{\varepsilon} f'_x(x,\ y) + g'_y(x,\ y)\right] dt < 0. \tag{1.6}$$

To prove this, we must find the dominant term of the integral in (1.6) for small ε, using the asymptotic approximations of the parts of Z_ε with error $O(\varepsilon \ln (1/\varepsilon))$ obtained in Chapter II; this term is negative. We do not reproduce the calculations since they are analogous to those in Secs. 3-6. By virtue of Poincaré's orbital stability criterion [30, 3, 55], it follows from (1.6) that the closed trajectory Z_ε is a stable limit cycle of (1.1).

Hence (1.1) has a unique closed trajectory having the zeroth asymptotic approximation Z_0. In fact, if there were more than one such trajectory, at least one of them would be unstable and this contradicts (1.6).

We conclude that, if ε is small enough, (1.1) has a unique periodic solution tending uniformly to a discontinuous periodic solution of (1.2) when $\varepsilon \to 0$. The rest of this chapter is devoted to the determination of asymptotic approximations for the cycle Z_ε and its period for $\varepsilon \to 0$.

2. Asymptotic Approximations for the Trajectory of a Periodic Solution

The limit cycle Z_ε is a closed curve in the (x,y) plane formed of a finite number of alternating slow-motion parts, junctions, fast-motion parts, and drop parts. To find asymptotic representations of the parts, we can apply the formulas obtained in Chapter II, using appropriate parts of the trajectory Z_0 as zeroth approximations in these formulas.

On Z_0, we consider successively the slow-motion part $\tilde{P}S$, the fast-motion part PS, and the slow-motion part $\hat{P}S$ (Fig. 48). In these sections we consider the points R^0, Q^-, R, Q^+, P^0, and $\hat{R}^0$, and on the part of the trajectory Z_ε close to the segment $R^0SP\hat{R}^0$, we consider the points A, B, D, E, and $\hat{A}$; the rules for selecting these points are described in Chapter II (Figs. 33, 35, 44, 45). The asymptotic determination of Z_ε is clearly equivalent to the asymptotic determination of $ABDE\hat{A}$.

The equation of the slow-motion part AB of the cycle Z_ε can be written as

$$x = x_Z(y, \varepsilon), \quad r_2^0 \operatorname{sign} g(S) \leqslant y \operatorname{sign} g(S) \leqslant r_2 \operatorname{sign} g(S); \tag{2.1}$$

asymptotic approximations of this function are obtained in Chapter II, Sec. 3.

The equation of the junction part BD of Z_ε can be written as

$$\eta = \eta_Z(\xi, \varepsilon), \quad -q \leqslant \xi \leqslant q \tag{2.2}$$

in the appropriate local coordinates in the neighborhood of S; asymptotic approximations of the function in (2.2) are obtained in Chapter II, Sec. 16, and they can be transferred to the coordinates x,y by using the formulas in Chapter II, Sec. 5.

The equation of the fast-motion part DE of Z_ε is

$$y = y_Z(x, \varepsilon),$$
$$q_1^+ \operatorname{sign} f_x''(S) \leqslant x \operatorname{sign} f_x''(S) \leqslant p_1^0 \operatorname{sign} f_x''(S); \tag{2.3}$$

asymptotic approximations for the function in (2.3) are obtained in Chapter II, Sec. 17.

The equation of the drop part $E\hat{A}$ of Z_ε is

$$x=\hat{x}_Z(y,\ \varepsilon),\qquad e\operatorname{sign} g(P)\leqslant y\operatorname{sign} g(P)\leqslant \hat{r}_2^0\operatorname{sign} g(P); \tag{2.4}$$

asymptotic approximations of the function in (2.4) are obtained in Chapter II, Sec. 20.

Hence the asymptotic formulas in Chapter II can be used for calculations near a discontinuous periodic solution of (1.1) with any required degree of accuracy, in particular for finding the amplitude of relaxation oscillations. The same formulas enable us to obtain asymptotic expansions for the period T_ε of relaxation oscillations of Z_ε. We must first calculate the time taken by the phase point of (1.1) to traverse the slow-motion part AB *(the slow-motion time)*, the junction part BD *(the junction time)*, the fast-motion part DE *(the fast-motion time)*, and the drop part $E\hat{A}$ *(the drop time)* of Z_ε.

3. Calculation of the Slow-Motion Time

Here we consider the slow-motion part AB of Z_ε, which is close to the segment R^0R of the stable part $x = x_0(y)$ of the curve Γ (Fig. 48); the equation of AB is given by (2.1). The time taken by the phase point of (1.1) to traverse AB will be calculated from the formula

$$T_{AB}=\int\limits_{AB}\frac{dy}{g(x,\ y)}=\int\limits_{r_2^0}^{r_2}\frac{dy}{g(x_Z(y,\ \varepsilon),\ y)}.$$

We replace the function (2.1) by its asymptotic representation [see (II.3.13)] and expand the integrand in powers of ε [see (II.3.7) and (II.6.7)]. Then, after some transformations and derivations of inequalities, we obtain the remainder term

$$T_{AB}=\int\limits_{r_2^0}^{r_2}\frac{dy}{g(x_0(y),\ y)}+\sum_{n=1}^{N-1}\varepsilon^n\int\limits_{r_2^0}^{r_2}\tilde{g}_n(y,\ x_0(y),\ x_1(y),\ \ldots,\ x_n(y))\,dy+O(\varepsilon^N), \tag{3.1}$$

where the $\tilde{g}_n(y, x_0(y), x_1(y), \ldots, x_n(y))$ are determined as the coefficients of the expansion under consideration.

It is clear that the first term in formula (3.1) is the time of traversal of the phase point of the degenerate system

(1.2) on the segment $R^\circ R$ of its trajectory Z_0 [cf. (1.4)]:

$$\int_{r_2^0}^{r_2} \frac{dy}{g(x_0(y), y)} = \int_{R^\circ R} \frac{dy}{g(x, y)} = T_{R^\circ R}. \tag{3.2}$$

We saw in Chapter II, Sec. 3 that each of the functions $x_i(y)$, $i = 1,2,\ldots$, can be expressed in terms of values of the right sides in (1.1) and certain of their derivatives on the arc $x = x_0(y)$ of Γ. Hence the integrand in the second term in (3.1) has the representation

$$\tilde{g}_n(y, x_0(y), x_1(y), \ldots, x_n(y)) \equiv g_n[f(x_0(y), y), g(x_0(y), y)],$$

and the $g_n[f(x,y),g(x,y)]$ can be effectively calculated for any positive integer n in terms of $f(x,y)$ and $g(x,y)$ and their derivatives.

We thus have the following asymptotic representation for the slow-motion time on AB:

$$T_{AB} = T_{R^\circ R} + \sum_{n=1}^{N-1} \varepsilon^n \int_{R^\circ R} g_n[f(x, y), g(x, y)]\, dy + O(\varepsilon^N), \tag{3.3}$$

where N is any positive integer. In particular, if we use (II.3.14), we find that

$$T_{AB} = T_{R^\circ R} + \varepsilon \int_{R^\circ R} \frac{f'_y(x, y)\, g'_x(x, y)}{f'^2_x(x, y)\, g(x, y)}\, dy + O(\varepsilon^2).$$

Hence, since

$$f'_x(x, y)\, dx + f'_y(x, y)\, dy = 0$$

on the curve $f(x,y) = 0$, we have the following first approximation for the slow-motion time on AB:

$$T_{AB} = T_{R^\circ R} + \varepsilon \int_{R^\circ R} \frac{-g'_x(x, y)}{f'_x(x, y)\, g(x, y)}\, dx + O(\varepsilon^2). \tag{3.4}$$

4. Calculation of the Junction Time

Consider the junction part BD of the trajectory Z_ε close to the segment Q^-SQ^+ of Z_0 (Fig. 48). In the neighborhood of S, we replace the coordinates x,y by local coordinates ξ,η (Chapter II, Sec. 5). In these coordinates, (1.1) has the form (II.5.6), and the equation of BD is given by (2.2).

It is clear that the junction time T_{BD} is equal to the traversal time of the phase point of the system (II.5.6) on the part (2.2), which can be calculated from the formula

$$T_{-q,\,q}=\varepsilon\int\limits_{-q}^{q}\frac{\gamma(\xi,\,\eta_Z(\xi,\,\varepsilon))}{\xi^2+\eta_Z(\xi,\,\varepsilon)}\,\delta(\xi,\;\eta_3(\xi,\;\varepsilon))\,d\xi \qquad (4.1)$$

[see the first equation in (II.5.6)]. Here we have used the notation in (II.6.3) and

$$\delta(\xi,\;\eta)\equiv\frac{1}{\beta(\xi,\;\eta)}; \qquad (4.2)$$

the function $\delta(\xi,\eta)$ is defined and strictly positive everywhere in U_0 (Chapter II, Sec. 5), so that, in particular, $\delta = \delta(0,\,0) > 0$.

Since the asymptotic representation of the function $\eta = \eta_Z(\xi,\varepsilon)$ is different on different sections of the junction part, we write the integral in (4.1) as the sum of four integrals on the intervals $-q \le \xi \le -\sigma_1$, $-\sigma_2 \le \xi \le 0$, $0 \le \xi \le \sigma_2$, and $\sigma_2 \le \xi \le q$. Here the quantities σ_1 and σ_2 are determined by the rule described in Chapter II, Sec. 16 (where we must put $a = N$).

We first calculate the time $T_{-q,-\sigma_1}$ for the motion of the phase point of the system (II.5.6) along the segment of the part (2.2) corresponding to the variation of ξ in the interval $-q \le \xi \le -\sigma_1$. Using Eq. (II.6.2) and the asymptotic approximation of the function (2.2) for the initial section of the junction part [see (II.6.12)], integrating by parts, and estimating the remainder term, we obtain

$$T_{-q,\,-\sigma_1}=-\int\limits_{-q}^{-\sigma_1}2\xi\delta(\xi,\;-\xi^2)\,d\xi-$$

$$-\sum_{n=1}^{N_1-1}\varepsilon^n\int\limits_{-q}^{-\sigma_1}\sum_{k=1}^{n}\frac{\delta'^{\,(k-1)}_{\xi\,\eta}(\xi,-\xi^2)}{k!}\sum_{\substack{i_1+\ldots+i_k=n,\\ i_j\geqslant 1}}\eta_{i_1}(\xi)\ldots\eta_{i_k}(\xi)\,d\xi+$$

$$+\sum_{n=1}^{N_1-1}\varepsilon^n\sum_{k=1}^{n}\frac{\delta_\eta^{(k-1)}(\xi,\;-\xi^2)}{k!}\sum_{\substack{i_1+\ldots+i_k=n,\\ i_j\geqslant 1}}\eta_{i_1}(\xi)\ldots\eta_{i_k}(\xi)\Big|_{-q}^{-\sigma_1}+O(\varepsilon^N). \qquad (4.3)$$

Starting from the asymptotic expansions (II.7.2), we can verify that the functions $\eta_i(\xi)$, $i = 1,2,\ldots$, satisfy the relation

$$\frac{\delta_{\xi\eta}^{\prime\,(k-1)}(\xi,-\xi^2)}{k!}\sum_{\substack{i_1+\ldots+i_k=n,\\ i_j\geqslant 1}}\eta_{i_1}(\xi)\ \ldots\ \eta_{i_k}(\xi)=\frac{1}{\xi^{3n-2k}}\sum_{\nu=0}^{\infty}\eta_\nu^{n,k}\xi^\nu,\quad \xi\to-0.$$

By using these expansions, we can regularize the integrals in the second term of (4.3) [see (II.13.14)]:

$$\sum_{n=1}^{N_1-1}\varepsilon^n\int_{-q}^{-\sigma_1}\sum_{k=1}^{n}\frac{\delta_{\xi}^{\prime}\eta^{(k-1)}(\xi,-\xi^2)}{k!}\sum_{\substack{i_1+\ldots+i_k=n,\\ i_j\geqslant 1}}\eta_{i_1}(\xi)\ldots\eta_{i_k}(\xi)\,d\xi=$$

$$=\sum_{n=1}^{N-1}\varepsilon^n\sum_{k=1}^{n}\oint_{-q}^{0}\frac{\delta_{\xi}^{\prime}\eta^{(k-1)}(\xi,-\xi^2)}{k!}\sum_{\substack{i_1+\ldots+i_k=n,\\ i_j\geqslant 1}}\eta_{i_1}(\xi)\ldots\eta_{i_k}(\xi)\,d\xi+$$
$$+\mathfrak{S}_1(-\sigma_1)+O(\varepsilon^N). \tag{4.4}$$

An expression for $\mathfrak{S}_1(-\sigma_1)$ can be obtained from Chapter II, Sec. 13 and writing out in detail the corresponding asymptotic expansions of the integrals [cf. (II.13.1)-(II.13.3)]. We do not reproduce this expression in detail; we only note that it is a sum of terms of the form

$$c\varepsilon^n\ln\sigma_1,\quad 1\leqslant n\leqslant N;$$
$$c\varepsilon^n\sigma_1^\nu,\quad 1\leqslant n\leqslant N_1-1,\quad -3n-3\leqslant\nu\leqslant\left]\frac{N-n}{\lambda_1}\right[,\quad \nu\neq 0,$$

arising in the regularization, and depending strictly on σ_1; the remaining terms arising in the regularization are included in the remainder term.

We write the third term in (4.3) as

$$\sum_{n=1}^{N_1-1}\varepsilon^n\sum_{k=1}^{n}\frac{\delta_\eta^{\prime(k-1)}(\xi,-\xi^2)}{k!}\sum_{\substack{i_1+\ldots+i_k=n,\\ i_j\geqslant 1}}\eta_{i_1}(\xi)\ldots\eta_{i_k}(\xi)\Bigg|_{-q}^{-\sigma_1}=$$
$$=\mathfrak{S}_2(-\sigma_1)-\mathfrak{S}_2(-q), \tag{4.5}$$

where $\mathfrak{S}_2(-\sigma_1)$ and $\mathfrak{S}_2(-q)$ are determined in the obvious fashion.

We shall also calculate the time $T_{-\sigma_1,0}$ for the motion of the phase point of the system (II.5.6) along the section of the part (2.2) corresponding to the interval $-\sigma_1\leq\xi\leq 0$ of values of ξ. Using Eq. (II.10.2) and the asymptotic approximation of the function (2.2) in the left part of a small (together with ε) neighborhood of the junction point [see (II.10.16) and (II.12.1)], and performing some elementary transformations, we find that

$$T_{-\sigma_1,0} = -\int_{-\sigma_1}^{0} 2\xi\delta(\xi, -\xi^2)\,d\xi -$$

$$-\sum_{n=2}^{N_2+2}\mu^n \int_{-\sigma_1}^{0} \sum_{k=1}^{[n/2]} \frac{\delta'_{\xi}{}_{\eta}^{(k-1)}(\xi,-\xi^2)}{k!} \sum_{\substack{i_1+\ldots+i_k=n,\\ i_j\geqslant 2}} v_{i_1-2}\left(\frac{\xi}{\mu}\right)\ldots v_{i_k-2}\left(\frac{\xi}{\mu}\right)d\xi +$$

$$+\sum_{n=2}^{N_2+2}\mu^n \sum_{k=1}^{[n/2]} \frac{\delta_{\eta}^{(k-1)}(\xi,-\xi^2)}{k!} \sum_{\substack{i_1+\ldots+i_k=n,\\ i_j\geqslant 2}} v_{i_1-2}\left(\frac{\xi}{\mu}\right)\ldots v_{i_k-2}\left(\frac{\xi}{\mu}\right)\Bigg|_{-\sigma_1}^{0} + O(\varepsilon^N); \tag{4.6}$$

here the expression $v_0(u)$, occurring for $i_j = 2$, denotes the function $z_0(u)$ (Chapter II, Sec. 12).

Now expanding the function $\delta'_{\xi}{}_{\eta}^{(k-1)}(\xi,-\xi^2)$ in a Maclaurin series and putting $\xi = \xi u$ (we recall that, when $\varepsilon \to 0$, $\omega_1 \to \infty$, but $\mu\omega_1 = \sigma_1 \to 0$), we find that

$$\frac{1}{k!}\delta'_{\xi}{}_{\eta}^{(k-1)}(\mu u, -\mu^2u^2) = \sum_{\nu=0}^{\infty}\mu^{\nu}\delta_{\nu}^{k}u^{\nu}, \quad \mu u \to 0; \tag{4.7}$$

the coefficients δ_{ν}^{k} can be effectively calculated. If we make the change $\xi = \mu u$ in the variable of integration and use (4.7), we find that the second term in (4.6) can be written as

$$\sum_{n=3}^{N_2+3}\mu^n \sum_{m=2}^{n-1}\sum_{k=1}^{[m/2]} \delta_{n-m-1}^{k}\int_{-\omega_1}^{0} u^{n-m-1} \sum_{\substack{i_1+\ldots+i_k=m,\\ i_j\geqslant 2}} v_{i_1-2}(u)\ldots v_{i_k-2}(u)\,du + O(\varepsilon^N).$$

The integrals here can be regularized [see (II.13.21)] by first using relation (II.12.12) to find the asymptotic expansion of the integrand for $u \to -\infty$. We thus obtain the following expression for the second term in (4.6):

$$\sum_{n=2}^{N_2+2}\mu^n \int_{-\sigma_1}^{0} \sum_{k=1}^{[n/2]} \frac{\delta'_{\xi}{}_{\eta}^{(k-1)}(\xi,-\xi^2)}{k!} \sum_{\substack{i_1+\ldots+i_k=n,\\ i_j\geqslant 2}} v_{i_1-2}\left(\frac{\xi}{\mu}\right)\ldots v_{i_k-2}\left(\frac{\xi}{\mu}\right)d\xi =$$

$$= \sum_{n=3}^{3N-1}\mu^n \sum_{m=2}^{n-1}\sum_{k=1}^{[m/2]} \delta_{n-m-1}^{k}\oint_{-\infty}^{0} u^{n-m-1} \sum_{\substack{i_1+\ldots+i_k=m,\\ i_j\geqslant 2}} v_{i_1-2}(u)\ldots v_{i_k-2}(u)\,du +$$

$$+\sum_{n=1}^{N}\mathcal{A}_n\mu^{3n}\ln\frac{1}{\mu} + \mathfrak{S}_3(-\sigma_1) + O(\varepsilon^N). \tag{4.8}$$

An expression for $\mathfrak{S}_3(-\sigma_1)$ is obtained from Chapter II, Sec. 13 and writing in detail the asymptotic expansions of the integrals [cf. (II.13.20)]. This expression is a sum of terms of the form $c\mu^n \ln \sigma_1$ and $c\mu^n\sigma_1^\nu$ arising in the regularization, depending explicitly on σ_1, and of order lower than ε. Terms arising in the regularization that do not depend on $O(\varepsilon^N)$ occur only in terms of the form σ_1 in the asymptotic expansions of the integrands. These terms c/μ are written separately in (4.8); it is easily seen that their coefficients can be effectively calculated:

$$\mathcal{A}_n = -\sum_{m=2}^{3n-1}\sum_{k=1}^{K}\delta^k_{3n-m-1}a^{m,k}_{n-k}, \quad K = \min\left\{\left[\frac{m}{2}\right], n\right\}.$$

The third term in (4.6) can be written as follows in the natural notation:

$$\sum_{n=2}^{N_2+2}\mu^n\sum_{k=1}^{[n/2]}\frac{\delta_\eta^{(k-1)}(\xi, -\xi^2)}{k!}\sum_{\substack{i_1+\ldots+i_k=n,\\ i_j\geqslant 2}} v_{i_1-2}\left(\frac{\xi}{\mu}\right)\ldots v_{i_k-2}\left(\frac{\xi}{\mu}\right)\Big|^0_{-\sigma_1} =$$

$$= \mathfrak{S}_4(0) - \mathfrak{S}_4(-\sigma_1). \tag{4.9}$$

To find an asymptotic formula for the time $T_{-q,0}$ of motion on the first half of the junction part, it only remains to combine (4.3) and (4.6) and use (4.4), (4.5), (4.8), and (4.9). Here it is essential that all terms depending explicitly on σ_1 vanish, since

$$\begin{aligned}\mathfrak{S}_2(-\sigma_1) - \mathfrak{S}_4(-\sigma_1) = O(\varepsilon^N),\\ \mathfrak{S}_1(-\sigma_1) + \mathfrak{S}_3(-\sigma_1) = O(\varepsilon^N).\end{aligned} \tag{4.10}$$

As an example we prove the first of relations (4.10). It follows from (II.7.2) that

$$\mathfrak{S}_2(-\sigma_1) = \sum_{n=1}^{N_1-1}\varepsilon^n\sum_{k=1}^{n}\frac{\delta_\eta^{(k-1)}(-\sigma_1, -\sigma_1^2)}{k!}\sum_{\substack{i_1+\ldots+i_k=n,\\ i_j\geqslant 1}}\eta_{i_1}(-\sigma_1)\ldots\eta_{i_k}(-\sigma_1) =$$

$$= \sum_{n=1}^{N_1-1}\varepsilon^n\sum_{k}^{n}\frac{\delta_\eta^{(k-1)}(-\sigma_1, -\sigma_1^2)}{k!}\times$$

$$\times\sum_{\nu=0}^{N+3n-2k}(-1)^{\nu-3n}\sigma_1^{\nu+2k-3n}\sum_{\substack{i_1+\ldots+i_k=\nu,\\ j_1+\ldots+j_k=n}}\eta^{j_1}_{i_1}\ldots\eta^{j_k}_{i_k} + O(\varepsilon^N);$$

but, by virtue of the asymptotic expansion (II.12.9),

$$\mathfrak{S}_4(-\sigma_1) = \sum_{n=2}^{N_2+2} \mu^n \sum_{k=1}^{[n/2]} \frac{\delta_\eta^{(k-1)}(-\sigma_1, -\sigma_1^2)}{k!} \times$$
$$\times \sum_{\substack{i_1+\ldots+i_k=n,\\ i_j \geqslant 2}} v_{i_1-2}\left(-\frac{\sigma_1}{\mu}\right)\ldots v_{i_k-2}\left(-\frac{\sigma_1}{\mu}\right) =$$
$$= \sum_{n=1}^{N_1-1} \varepsilon^n \sum_{k=1}^{n} \frac{\delta_\eta^{(k-1)}(-\sigma_1, -\sigma_1^2)}{k!} \times$$
$$\times \sum_{\nu=0}^{N+3n-2k} (-1)^{\nu-3n} \sigma_1^{\nu+2k-3n} \gamma^n \sum_{\substack{i_1+\ldots+i_k=\\ =n-k,\\ j_1+\ldots+j_k=\nu}} a_{i_1}^{j_1}\ldots a_{i_k}^{j_k} + O(\varepsilon^N).$$

It is thus sufficient to prove that

$$\gamma^n \sum_{\substack{i_1+\ldots+i_k=n-k,\\ j_1+\ldots+j_k=\nu}} a_{i_1}^{j_1}\ldots a_{i_k}^{j_k} = \sum_{\substack{i_1+\ldots+i_k=\nu,\\ j_1+\ldots+j_k=n}} \eta_{i_1}^{j_1}\ldots\eta_{i_k}^{j_k}$$

for all possible values of the indices. But this is a direct consequence of (II.12.11).

We thus have the following asymptotic representation for the time of motion along the first half of the junction part BD:

$$T_{-q,0} = -\int_{-q}^{0} 2\xi\delta(\xi, -\xi^2)\,d\xi + \sum_{n=3}^{3N-1} A_n^{(1)}\varepsilon^{n/3} + \sum_{n=1}^{N} A_n^{(2)}\varepsilon^n \ln\frac{1}{\varepsilon} - \mathfrak{S}_2(-q) + \mathfrak{S}_4(0) + O(\varepsilon^N). \tag{4.11}$$

In particular, if we use formulas (II.16.12) and (II.6.9), we obtain

$$T_{-q,0} = -\int_{-q}^{0} 2\xi\delta(\xi, -\xi^2)\,d\xi + \gamma^{2/3}\delta z_0(0)\,\varepsilon^{2/3} - \frac{1}{6}\gamma\delta'_\xi\,\varepsilon\ln\frac{1}{\varepsilon} +$$
$$+ \left\{\frac{1}{6}\gamma\delta'_\xi \ln\gamma + \oint_{-q}^{0} \frac{\gamma(\xi, -\xi^2)\,\delta'_\xi(\xi, -\xi^2)}{2\xi}\,d\xi - \gamma\delta'_\xi \oint_{-\infty}^{0} z_0(u)\,du\right\}\varepsilon +$$
$$+ \frac{\alpha(-q, -q^2)}{-2q}\,\varepsilon + \alpha v_1(0)\,\varepsilon + O(\varepsilon^{4/3}). \tag{4.12}$$

We now calculate the time T_{0,σ_2} taken by the phase point of the system (II.5.6) to traverse the section of the part (2.2) corresponding to the interval $0 \leq \xi \leq \sigma_2$. Using Eq. (II.10.2) and the asymptotic approximation of the function (2.2) in the right side of a neighborhood of the junction point, small when ε is small [see (II.10.17)], carrying out some transformations, and estimating the remainder term, we find that

$$T_{0,\sigma_2} = -\sum_{n=2}^{N_3+2} \int_0^{\sigma_2} \sum^{[n/2]} \frac{\delta'^{(k-1)}_{\xi\eta}(\xi, 0)}{k!} \sum_{\substack{i_1+\ldots+i_k=n,\\ i_j \geq 2}} v_{i_1-2}\left(\frac{\xi}{\mu}\right)\ldots v_{i_k-2}\left(\frac{\xi}{\mu}\right) d\xi +$$

$$+ \sum_{n=2}^{N_3+2} \mu^n \sum_{k=1}^{[n/2]} \frac{\delta^{(k-1)}_{\eta}(\xi, 0)}{k!} \sum_{\substack{i_1+\ldots+i_k=n,\\ i_j \geq 2}} v_{i_1-2}\left(\frac{\xi}{\mu}\right)\ldots v_{i_k-2}\left(\frac{\xi}{\mu}\right)\Big|_0^{\sigma_2} + O(\varepsilon^N) \tag{4.13}$$

[cf. (4.6)]; here $v_0(u)$ is the function introduced in Chapter II, Sec. 9.

Now employing the expansion

$$\frac{1}{k!}\delta'^{(k-1)}_{\xi\eta}(\mu u, 0) = \sum_{\nu=0}^{\infty} \mu^\nu \tilde{\delta}^k_\nu u^\nu, \quad \mu u \to 0$$

[cf. (4.7)] whose coefficients can be effectively calculated, and making the change $\xi = \mu u$ in the variable of integration, we can write the first term in (4.13) as

$$\sum_{n=3}^{N_3+3} \mu^n \sum_{m=2}^{n-1} \sum_{k=1}^{[m/2]} \tilde{\delta}^k_{n-m-1} \int_0^{\omega_2} u^{n-m-1} \sum_{\substack{i_1+\ldots+i_k=m,\\ i_j\geq 2}} v_{i_1-2}(u)\ldots v_{i_k-2}(u)\,du + O(\varepsilon^N).$$

Starting from (II.12.13), we find the asymptotic expansion of the integrand for $u \to +\infty$, and can then regularize the relevant integrals. As a result, the first term in (4.13) is transformed as follows:

$$\sum_{n=2}^{N_3+2} \mu^n \int_0^{\sigma_2} \sum_{k=1}^{[n/2]} \frac{\delta'^{(k-1)}_{\xi\eta}(\xi, 0)}{k!} \sum_{\substack{i_1+\ldots+i_k=n,\\ i_j\geq 2}} v_{i_1-2}\left(\frac{\xi}{\mu}\right)\ldots v_{i_k-2}\left(\frac{\xi}{\mu}\right) d\xi =$$

$$= \sum_{n=3}^{3N-1} \mu^n \sum_{m=2}^{n-1} \sum_{k=1}^{[m/2]} \tilde{\delta}^k_{n-m-1} \oint_0^{\infty} u^{n-m-1} \sum_{\substack{i_1+\ldots+i_k=m,\\ i_j\geq 2}} v_{i_1-2}(u)\ldots v_{i_k-2}(u)\,du +$$

$$+ \sum_{n=3}^{3N} \mu^n \sum_{\nu=0}^{\pi(n-2)} \mathcal{A}^*_{n,\nu} \ln^\nu \frac{1}{\mu} + \mathfrak{S}_5(\sigma_2) + O(\varepsilon^N). \tag{4.14}$$

The expression $\mathfrak{S}_5(\sigma_2)$ represents a sum of terms of the form $c\mu^n\sigma_2^m \ln^\nu (1/\mu) \ln^\kappa \sigma_2$ arising in the regularization, depending explicitly on σ_2, and of lower order than $O(\varepsilon^N)$. Terms not depending on σ_2 arise in the regularization only from terms of the form $cu^{-1}\ln^\nu u$ in the asymptotic expansion of the integrands. These terms $cu^{-1}\ln^\nu u$ are written separately in (4.14); their coefficients $\mathcal{A}^*_{n,\nu}$ can be effectively determined by writing in detail the appropriate asymptotic expansions (Chapter II, Sec. 13).

Now consider the second term in (4.13). Making the variable change $\xi = \mu u$ in the integral and using the Maclaurin expansion

$$\frac{1}{k!}\delta_\eta^{(k-1)}(\mu u, 0) = \sum_{\nu=0}^{\infty} \mu^\nu \tilde{\tilde{\delta}}_\nu^k u^\nu, \quad \mu u \to 0,$$

we obtain

$$\sum_{n=2}^{N_3+2} \mu^n \sum_{m=2}^{n} \sum_{k=1}^{[m/2]} \tilde{\tilde{\delta}}_{n-m}^k u^{n-m} \sum_{\substack{i_1+\ldots+i_k=m, \\ i_j\geqslant 2}} v_{i_1-2}(u)\ldots v_{i_k-2}(u) + O(\varepsilon^N).$$

Analysis of this expression and the use of (II.12.13) yield the asymptotic expansion for $u \to +\infty$, in particular the terms of the form $c \ln^\nu u$ in the expansion. It is clear that after the substitution $u = \omega_2 = \sigma_2/\mu$, only these terms lead to terms not depending on σ_2. When the necessary calculations have been carried out we find that the second term in (4.13) has the following representation:

$$\sum_{n=2}^{N_3+2} \mu^n \sum_{k=1}^{[n/2]} \frac{\delta_\eta^{(k-1)}(\xi, 0)}{k!} \sum_{\substack{i_1+\ldots+i_k=n, \\ i_j\geqslant 2}} v_{i_1-2}\left(\frac{\xi}{\mu}\right) \ldots v_{i_k-2}\left(\frac{\xi}{\mu}\right)\Bigg|_0^{\sigma_2} =$$

$$= \sum_{n=2}^{3N} \mu^n \sum_{\nu=0}^{\pi(n-2)} \mathcal{A}^{**}_{n,\nu} \ln^\nu \frac{1}{\mu} + \mathfrak{S}_6(\sigma_2) - \mathfrak{S}_7(0) + O(\varepsilon^N); \qquad (4.15)$$

the coefficients $\mathcal{A}^{**}_{n,\nu}$ can be effectively determined, and the meaning of $\mathfrak{S}_6(\sigma_2)$ and $\mathfrak{S}_7(0)$ is clear.

Finally we investigate the time $T_{\sigma_2,q}$ for the phase point of the system (II.5.6) to traverse the section of the part (2.2) corresponding to the interval $\sigma_2 \leq \xi \leq q$. Using (4.2) and the asymptotic expansion of the function (2.2) at the end of the junction part [see (II.14.13)], we obtain

$$T_{\sigma_2,q}=\varepsilon\int_{\sigma_2}^{q}\frac{\alpha(\xi,\eta_3(\xi,\varepsilon))}{\xi^2+\eta_3(\xi,\varepsilon)}d\xi=\frac{1}{\gamma}\mu^3\int_{\sigma_2}^{q}\frac{\alpha\left(\xi,\sum\limits_{n=2}^{N_4+2}\mu^n\sum\limits_{\nu=0}^{\pi(n-2)}\zeta_{n,\nu}(\xi)\ln^\nu(1/\mu)\right)}{\xi^2+\sum\limits_{n=2}^{N_4+2}\mu^n\sum\limits_{\nu=0}^{\pi(n-2)}\zeta_{n,\nu}(\xi)\ln^\nu(1/\mu)}d\xi+O(\varepsilon^N).$$

The expansion of the integrand in this formula in terms of the quantities $\mu^n\ln^\nu(1/\mu)$ is realized with the same accuracy as in the analogous transformation in Chapter II, Sec. 14. If we now use the asymptotic expansions (II.15.3) for the $\zeta_{n,\nu}(\xi)$, regularize the resulting integrals [see (II.13.17)], and carry out some elementary transformations, we obtain

$$T_{\sigma_2,q}=\mu^3\oint_0^q\frac{\alpha(\xi,0)}{\gamma\xi^2}d\xi+\sum_{n=5}^{3N}\mu^n\sum_{\nu=0}^{\pi(n-2)-1}\ln^\nu\frac{1}{\mu}\times$$

$$\times\sum_{k=1}^{\left[\frac{n-3}{2}\right]}\sum_{m=0}^{k}\oint_0^q\frac{(-1)^m\alpha_\eta^{(k-m)}(\xi,0)}{(k-m)!\,\gamma\xi^{2m+2}}W_{n,k}^{\nu}(\xi)\,d\xi+\mathfrak{S}_8(\sigma_2)+O(\varepsilon^N)\qquad(4.16)$$

[see (II.14.9)]. Here $\mathfrak{S}_8(\sigma_2)$ denotes the sum of terms of the form $c\mu^n\sigma_2^m\ln^\nu(1/\mu)\ln^\kappa\sigma_2$ arising in the regularization, depending strictly on σ_2, and of order lower than $O(\varepsilon^N)$.

To find an asymptotic formula for the time $T_{0,q}$ for motion on the second half of the junction part, it remains to combine (4.13) and (4.16) and use (4.14) and (4.15). Here it is essential that all terms depending explicitly on σ_2 vanish, since

$$\mathfrak{S}_8(\sigma_2)+\mathfrak{S}_6(\sigma_2)-\mathfrak{S}_5(\sigma_2)=O(\varepsilon^N)\qquad(4.17)$$

[cf. (4.10)]. Relation (4.17) can be proved by direct elementary methods, but the verification is long and difficult; the method of Chapter II, Sec. 13 is used, in which the appropriate asymptotic expansions are written in detail and formulas (II.15.5) are used.

We thus have the following asymptotic representation of the traversal time on the second half of BD:

$$T_{0,q}=\sum_{n=2}^{3N}\varepsilon^{n/3}\sum_{\nu=0}^{\pi(n-2)}A_{n,\nu}^{*}\ln^\nu\frac{1}{\varepsilon}-\mathfrak{S}_7(0)+O(\varepsilon^N).\qquad(4.18)$$

In particular, if we use (II.16.12), we find that

$$T_{0,q}=\gamma^{2/3}\delta(\Omega_0-v_0(0))\varepsilon^{2/3}+\frac{1}{3}\alpha'_\xi\varepsilon\ln\frac{1}{\varepsilon}+$$
$$+\left\{\alpha\Omega_1-\frac{1}{3}\alpha'_\xi\ln\gamma-\gamma\delta'_\xi-\gamma\delta'_\xi\oint_0^\infty v_0(u)\,du+\oint_0^q\frac{\alpha(\xi,0)}{\xi^2}\,d\xi\right\}\varepsilon-$$
$$-\alpha v_1(0)\,\varepsilon+O(\varepsilon^{4/3}). \qquad (4.19)$$

Lemma 1. The time taken by the phase point of the nondegenerate system (1.1) to traverse the junction part BD of the trajectory Z_ε close to the segment Q^-SQ^+ of the trajectory Z_0 has the asymptotic representation

$$T_{BD}=T_{Q^-S}+\sum_{n=2}^{N+1}\varepsilon^{n/3}\sum_{\nu=0}^{\pi(n-2)}A_{n,\nu}\ln^\nu\frac{1}{\varepsilon}+O\left(\varepsilon^{\frac{N+2}{3}}\ln^{\pi(N)}\frac{1}{\varepsilon}\right), \qquad (4.20)$$

where N is any positive integer. Integration of the system (1.1) is not necessary in the calculation of the coefficients $A_{n,\nu}$.

Since $T_{BD} = T_{-q,q} = T_{-q,0} + T_{0,q}$ [see (4.1)], the assertion of the lemma follows from (4.11) and (4.18). In fact (II.5.3) and (II.5.7) imply that

$$-\int_{-q}^{0}2\xi\delta(\xi,-\xi^2)\,d\xi=\int_{q_2^-}^{s_2}\frac{dy}{g(x_0(y),y)}=\int_{Q^-S}\frac{dy}{g(x,y)}=T_{Q^-S},$$

where T_{Q^-S} is the time taken by the phase point of the degenerate system (1.2) to traverse the part Q^-S of Z_0. It is clear that $\mathfrak{S}_4(0)=\mathfrak{S}_7(0)$ [see (4.9) and (4.15)]. Finally, (4.5) implies that $\mathfrak{S}_2(-q)$ is a polynomial in ε.

The coefficients $A_{n,\nu}$ in (4.20) are calculated directly in terms of values of the right sides of (1.1) and certain of their derivatives on the segment Q^-SQ^+ of Z_0; this is easily seen by analyzing the determination of the coefficients in (4.4), (4.8), (4.14), (4.15), and (4.16).

We saw in Chapter II, Sec. 6 that each of the functions $\eta_i(\xi)$, $i = 1,2,\ldots$, can be expressed in terms of values of $\gamma(\xi,\eta)$ and some of its derivatives on the curve $\eta = -\xi^2$ for $-q \leq \xi < 0$. Hence the generalized integral

$$\oint_{-q}^{0}\frac{\delta'^{(k-1)}_{\xi\eta}(\xi,-\xi^2)}{k!}\sum_{\substack{i_1+\ldots+i_k=n,\\ i_j\geqslant 1}}\eta_{i_1}(\xi)\ldots\eta_{i_k}(\xi)\,d\xi$$

[see (4.4)] is a generalized line integral on the part indicated. The change of variables (II.5.3) leads to a generalized line integral, on the segment $Q^{-}S$ of Γ, of a function that can be expressed in terms of only $f(x,y)$, $g(x,y)$, and their derivatives [cf. (II.13.15) and (II.13.16)]; the terms outside the integral are expressed in terms of values of these functions and their derivatives at the junction point S. A similar assertion holds concerning the generalized integrals in (4.16).

Furthermore, the generalized integral

$$\oint_{-\infty}^{0} u^{n-m-1} \sum_{\substack{i_1+\ldots+i_k=m,\\ i_j\geqslant 2}} v_{i_1-2}(u)\ldots v_{i_k-2}(u)\,du$$

[see (4.8)] is a linear combination of universal constants (generalized integrals of universal functions), with coefficients that can be expressed in terms of values of the right sides in (1.1) and their derivatives at the junction point S [cf. (II.12.17)]. A similar result holds concerning the generalized integrals in (4.14).

Finally, the constant δ^{k}_{n-m-1} [see (4.8)] is expressed in terms of the appropriate derivative of $g(x,y)$ at S [see (4.7)], and the constant $\mathcal{A}_n$ [see (4.8)] is expressed in terms of values of the right-hand sides of (1.1) and their derivatives at S and universal constants (coefficients in asymptotic expansions of universal functions). The situation is similar for the constants in (4.14) and (4.15).

As an example of the representation (4.20), we have the following formula [see (4.12) and (4.19)]:

$$\begin{aligned}
&T_{BD}=T_{Q-S}+A_{2,0}\varepsilon^{2/3}+A_{3,1}\varepsilon\ln\frac{1}{3}+A_{3,0}\varepsilon+O(\varepsilon^{4/3}),\\
&A_{2,0}=\gamma^{2/3}\delta\Omega_0,\quad A_{3,1}=\frac{1}{6}(\gamma\delta'_\xi+2\gamma'_\xi\delta),\\
&A_{3,0}=\gamma'_\xi\delta\mathbf{\Omega}_1-\frac{1}{6}(\gamma\delta'_\xi+2\gamma'_\xi\delta)\ln\gamma-\gamma\delta'_\xi\boldsymbol{I}_0+\\
&\quad+\oint_{-q}^{0}\frac{\gamma(\xi,-\xi^2)\,\delta'_\xi(\xi,-\xi^2)}{2\xi}\,d\xi+\oint_{0}^{q}\frac{\alpha(\xi,0)}{\xi^2}\,d\xi+\frac{\alpha(-q,-q^2)}{-2q}.
\end{aligned}\tag{4.21}$$

The quantities here, by virtue of formulas in Chapter II, Secs.

5, 12, and 13, can be directly expressed in terms of the right sides in (1.1), without integration of the system

$$\varphi'(0)=\sqrt{\left|\frac{f_x''(S)}{2f_y'(S)}\right|},$$

$$\varphi''(0)=\frac{f_x^{(3)}(S)\,f_y'(S)-3f_{xy}''(S)\,f_x''(S)}{6f_y'^2(S)}\sqrt{\left|\frac{2f_y'(S)}{f_x''(S)}\right|}\operatorname{sign}g(S);$$

$$\alpha=\sqrt{\frac{2}{|f_x''(S)\,f_y'(S)|}},\quad \beta=|g(S)|,$$

$$\gamma=|g(S)|\sqrt{\frac{2}{|f_x''(S)\,f_y'(S)|}},\qquad \delta=\frac{1}{|g(S)|};$$

$$\alpha_\xi'=-\frac{2f_x^{(3)}(S)}{3f_x''^2(S)},\quad \beta_\xi'=g_x'(S)\sqrt{\left|\frac{2f_y'(S)}{f_x''(S)}\right|}\operatorname{sign}f_y'(S),$$

$$\gamma_\xi'=\frac{6f_x''(S)\,g_x'(S)-2f_x^{(3)}(S)\,g(S)}{3f_x''^2(S)}\operatorname{sign}g(S),$$

$$\delta_\xi'=-\frac{g_x'(S)}{g^2(S)}\sqrt{\left|\frac{2f_y'(S)}{f_x''(S)}\right|}\operatorname{sign}f_y'(S); \tag{4.22}$$

$$\oint_{-q}^{0}\frac{\gamma(\xi,-\xi^2)\,\delta_\xi'(\xi,-\xi^2)}{2\xi}\,d\xi=\oint_{Q^-S}\frac{-g_x'(x,y)}{f_x''(x,y)\,g(x,y)}\,dx+\frac{g_x'(S)}{f_x''(S)\,g(S)}\ln\varphi'(0),$$

$$\oint_{0}^{q}\frac{\alpha(\xi,0)}{\xi^2}\,d\xi=\oint_{SQ+}\frac{dx}{f(x,y)}+\alpha\frac{\varphi''(0)}{2\varphi'^2(0)}\operatorname{sign}f_x''(S)+\alpha_\xi'\ln\varphi'(0);$$

$$\Omega_0=\lim_{u\to\infty}v_0(u),\quad \boldsymbol{\Omega}_1=\lim_{u\to\infty}[\boldsymbol{v}_1(u)-\ln u];$$

$$\boldsymbol{I}_0=1+\oint_{-\infty}^{0}z_0(u)\,du+\oint_{0}^{\infty}v_0(u)\,du;$$

$$\frac{\alpha(-q,-q^2)}{-2q}=\frac{1}{f_x'(Q-)}.$$

5. Calculation of the Fast-Motion Time

Consider the fast-motion part DE of the trajectory Z_ε close to the segment Q^+P^0 of the trajectory Z_0 (Fig. 48); the equation of DE is given by (2.3). The time taken by the phase point of (1.1) to traverse DE will be calculated from the formula

$$T_{DE}=\varepsilon\int_{DE}\frac{dx}{f(x,y)}=\varepsilon\int_{q_1^+}^{p_1^0}\frac{dx}{f(x,y_Z(x,\varepsilon))}.$$

We replace the function (2.3) by its asymptotic representation [see (II.17.14)], expand the integrand in a series of terms of the form $\varepsilon^{n/3}\ln^{\nu}(1/\varepsilon)$, and apply elementary transformations to obtain

$$T_{DE}=\varepsilon\int_{q_1^+}^{p_1^0}\frac{dx}{f(x,s_2)}+\sum_{n=5}^{N+4}\varepsilon^{n/3}\sum_{\nu=0}^{\pi(n-2)-1}\ln^{\nu}\frac{1}{\varepsilon}\times$$

$$\times\int_{q_1^+}^{p_1^0}\tilde{f}_{n,\nu}(x,\,y_{0,0}(x),\,y_{2,0}(x),\,\ldots,\,y_{n-5,\varkappa}(x))\,dx+O\left(\varepsilon^{\frac{N+5}{3}}\ln^{\pi(N)}\frac{1}{\varepsilon}\right),\tag{5.1}$$

where the functions $\tilde{f}_{n,\nu}(x,y_{0,0}(x),\ y_{2,0}(x),\ldots,y_{n-5,\kappa}(x))$ are the coefficients in the expansion.

We saw in Chapter II, Sec. 17 that each of the functions $y_{n,\nu}(x)$ can be expressed in terms of values of the right sides in (1.1) and their derivatives on the part SP of Z_0. Hence the integrand in the second term in (5.1) has the representation

$$\tilde{f}_{n,\nu}(x,\,y_{0,0}(x),\,y_{2,0}(x),\,\ldots,\,y_{n-5,\varkappa}(x))=f_{n,\nu}[f(x,s_2),\ g(x,s_2)],$$

and $f_{n,\nu}[f(x,y),g(x,y)]$ can be effectively calculated for $n\geq 5$ and ν, $0\leq\nu\leq\pi(n-2)-1$; it contains only the functions $f(x,y)$, $g(x,y)$, and their derivatives.

We thus have the following asymptotic representation for the fast-motion time on DE:

$$T_{DE}=\varepsilon\int_{Q^+P^0}\frac{dx}{f(x,y)}+\sum_{n=5}^{N+4}\varepsilon^{n/3}\sum_{\nu=0}^{\pi(n-2)-1}\ln^{\nu}\frac{1}{\varepsilon}\times$$

$$\times\int_{Q^+P^0}f_{n,\nu}[f(x,y),\ g(x,y)]\,dx+O\left(\varepsilon^{\frac{N+5}{3}}\ln^{\pi(N)}\frac{1}{\varepsilon}\right);\tag{5.2}$$

here N is any positive integer. In particular, it is obvious that

$$T_{DE}=\varepsilon\int_{Q^+P^0}\frac{dx}{f(x,y)}+O(\varepsilon^{5/3}).\tag{5.3}$$

6. Calculation of the Drop Time

In this section we investigate the drop part $E\hat{A}$ of the trajectory Z_ε close to the segment $P^0P\hat{R}^0$ of the trajectory Z_0 (Fig. 48); the equation of $E\hat{A}$ is given by (2.4). The drop time will be calculated from the formula

$$T_{E\hat{A}} = \int\limits_{E\hat{A}} \frac{dy}{g(x, y)}. \tag{6.1}$$

Since the asymptotic representations of the curve (2.4) are different on different sections of the drop part, we separate the integral in (6.1) into two integrals, one on EA and one on $G\hat{A}$. We recall that the choice of the point $G(\boldsymbol{x}, \boldsymbol{y})$ on the section $E\hat{A}$ of Z_ε depends on the accuracy required of the asymptotic representation of the drop part (Chapter II, Sec. 20).

We start by finding the traversal time of the phase point of (1.1) on EG. If the parametric form (II.19.11) of the equation of this arc is used, we can write

$$T_{EG} = \int\limits_{EG} \frac{dy}{g(x, y)} = \varepsilon \int\limits_1^{\omega} \frac{b\, dw}{wg\,(p_1 + sz_Z(w, \varepsilon),\ e + \varepsilon b \ln w)};$$

here ω is the value of w corresponding to the point G. By using the asymptotic representation of the curve (2.4) on the initial drop part (II.2-.17), expanding the integrand in a series of terms of the form $\varepsilon^{n/3} \ln^{\nu} (1/\varepsilon)$ (II.20.5), and making some elementary transformations and estimates, we arrive at the following relation:

$$T_{EG} = \varepsilon \int\limits_1^{\omega} \frac{b\, dw}{wg\,(p_1 + sz_{0,0}(w),\ p_2)} + \sum_{n=5}^{N+5} \varepsilon^{n/3} \sum_{\nu=0}^{\pi(n-2)-1} \ln^{\nu} \frac{1}{\varepsilon} \times$$

$$\times \int\limits_1^{\omega} g^*_{n,\nu}(p_1 + sz_{0,0}(w),\ z_{2,0}(w),\ \ldots,\ z_{n-5,\varkappa}(w))\, dw +$$

$$+ O\left(\varepsilon^{\frac{N+6}{3}} \ln^{\pi(N+1)} \frac{1}{\varepsilon}\right); \tag{6.2}$$

here the functions $g^*_{n,\nu}(p_1 + sz_{0,0}(w), z_{2,0}(w), \ldots, z_{n-5,\kappa}(w))$ are the coefficients of the expansion used, and $\omega = \varepsilon^{-(N+3)/3}$.

It follows from the asymptotic representation (II.21.1) of $z_{0,0}(w)$ that

$$\frac{b}{wg\,(p_1 + sz_{0,0}(w),\ p_2)} = \frac{b}{g(P)w} + O\left(\frac{1}{w^2}\right), \quad w \to +\infty.$$

Hence we can regularize the integral in the first term of (6.2) [cf. (II.13.18)-(II.13.21)]; using $\omega = \varepsilon^{-(N+3)/3}$ and relations (II.19.8), we obtain

$$\varepsilon \int_1^{\omega} \frac{b\,dw}{wg(p_1+sz_{0,0}(w),\ p_2)} =$$

$$= \varepsilon \oint_1^{\infty} \frac{b\,dw}{wg(p_1+sz_{0,0}(w),\ p_2)} - \frac{1}{f'_x(P)}\varepsilon \ln \omega + O\left(\varepsilon^{\frac{N+6}{3}}\right). \qquad (6.3)$$

The second term in (6.2) can be transformed similarly. By a calculation similar to that in Chapter II, Sec. 20 and the use of the expansion (II.21.5), we can derive the following asymptotic expansion [cf. (II.21.4)]:

$$g^*_{n,\nu}(p_1+sz_{0,0}(w),\ z_{2,0}(w),\ \ldots,\ z_{n-5,\varkappa}(w))^+ =$$

$$= \sum_{\alpha=0}^{\pi(n-2)-\nu-1} g_{\alpha}^{n,\nu}\frac{\ln^{\alpha} w}{w} + O\left(\frac{\ln^{\pi(n-2)-\nu} w}{w^2}\right).$$

These relations can be employed to regularize the integrals in the second term in (6.2):

$$\sum_{n=5}^{N+5} \varepsilon^{n/3} \sum_{\nu=0}^{\pi(n-2)-1} \ln^{\nu}\frac{1}{\varepsilon}\int_1^{\omega} g^*_{n,\nu}(p_1+sz_{0,0}(w),\ z_{2,0}(w),\ \ldots,\ z_{n-5,\varkappa}(w))\,dw =$$

$$= \sum_{n=5}^{N+5} \varepsilon^{n/3} \sum_{\nu=0}^{\pi(n-2)-1} \ln^{\nu}\frac{1}{\varepsilon}\oint_1^{\infty} g^*_{n,\nu}(p_1+sz_{0,0}(w),\ z_{2,0}(w),\ \ldots,\ z_{n-5,\varkappa}(w))\,dw +$$

$$+ \mathfrak{S}_0(\omega) + O\left(\varepsilon^{\frac{N+8}{3}}\ln\frac{1}{\varepsilon}\right). \qquad (6.4)$$

The term $\mathfrak{S}_0(\omega)$ is obtained by writing in detail the appropriate asymptotic expansions of the integrals, and is the sum of the terms arising in the regularization depending explicitly on ω:

$$\mathfrak{S}_0(\omega) = \sum_{n=5}^{N+5} \varepsilon^{n/3} \sum_{\nu=0}^{\pi(n-2)-1} \sum_{\beta=1}^{\pi(n-2)-\nu} \frac{g_{\beta-1}^{n,\nu}}{\beta} \ln^{\nu}\frac{1}{\varepsilon}\ln^{\beta}\omega.$$

We now find the time taken by the phase point of (1.1) to traverse $G\hat{A}$. By virtue of the equation (2.4) of this arc [cf. (II.19.5)],

$$T_{G\hat{A}} = \int_{G\hat{A}} \frac{dy}{g(x,y)} = \int_y^{\hat{r}_2^0} \frac{dy}{g(x_Z(y,\varepsilon),\ y)},$$

where $y = e + \varepsilon b \ln \omega$ is the ordinate of G (see Chapter II, Sec. 20). Using the asymptotic representation of the curve (2.4) at the end of the drop part (II.20.21) and applying the obvious transformations (Sec. 3), we obtain

$$T_{G\hat{A}} = \int_{y}^{\hat{r}_2^0} \frac{dy}{g(\hat{x}_0(y), y)} + \sum_{n=1}^{\left]\frac{N+3}{3}\right]} \varepsilon^n \times$$

$$\times \int_{y}^{\hat{r}_2^0} \tilde{g}_n(y, \hat{x}_0(y), \hat{x}_1(y), \ldots, \hat{x}_n(y))\,dy + O\left(\varepsilon^{\frac{N+3}{3}} \ln^{\pi(N+1)} \frac{1}{\varepsilon}\right).$$

Since the $x_i(y)$, $i = 0,1,\ldots$, are defined on the whole interval of variation of y between $\hat{s}_2^*$ and $\hat{s}_2$ (Chapter II, Secs. 3 and 19), this formula can be written as

$$T_{G\hat{A}} = \int_{p_2}^{\hat{r}_2^0} \frac{dy}{g(\hat{x}_0(y), y)} + \sum_{n=1}^{\left]\frac{N+3}{3}\right]} \varepsilon^n \int_{p_2}^{\hat{r}_2^0} \tilde{g}_n(y, \hat{x}_0(y), \ldots, \hat{x}_n(y))\,dy -$$

$$- \int_{p_2}^{y} \frac{dy}{g(\hat{x}_0(y), y)} - \sum_{n=1}^{\left]\frac{N+3}{3}\right]} \varepsilon^n \int_{p_2}^{y} \tilde{g}_n(y, \hat{x}_0(y), \ldots, \hat{x}_n(y))\,dy +$$

$$+ O\left(\varepsilon^{\frac{N+3}{3}} \ln^{\pi(N+1)} \frac{1}{\varepsilon}\right). \qquad (6.5)$$

By repeating the reasoning of Sec. 3, we find that the first two terms in (6.5) are transformed as follows:

$$\int_{p_2}^{\hat{r}_2^0} \frac{dy}{g(\hat{x}_0(y), y)} + \sum_{n=1}^{\left]\frac{N+3}{3}\right]} \varepsilon^n \int_{p_2}^{\hat{r}_2^0} \tilde{g}_n(y, \hat{x}_0(y), \ldots, \hat{x}_n(y))\,dy =$$

$$= T_{P\hat{R}^0} + \sum_{n=1}^{\left]\frac{N+3}{3}\right]} \varepsilon^n \int_{P\hat{R}^0} g_n[f(x, y), g(x, y)]\,dy; \qquad (6.6)$$

here $T_{P\hat{R}^0}$ is the time taken by the phase point of the degenerate system (1.2) to traverse the part $P\hat{R}^0$ of the trajectory Z_0 [see (3.2)].

Now consider the third term in (6.5). Using the relation

$$y = e + \varepsilon b \ln \omega = E_{N+2}(\varepsilon) + \varepsilon b \ln \omega + O\left(\varepsilon^{\frac{N+3}{3}} \ln^{\pi(N+1)} \frac{1}{\varepsilon}\right)$$

[see (II.20.21), (II.20.17), and (II.20.15)], and introducing the function

$$G_0(y) \equiv \frac{1}{g(\hat{x}_0(y),\, y)},$$

we conclude, by using Taylor's formula, that

$$\int_{p_2}^{y} \frac{dy}{g(\hat{x}_0(y),\, y)} = \sum_{k=1}^{\left]\frac{N+3}{2}\right]} \frac{[E_{N+2}(\varepsilon)+\varepsilon b \ln \omega - p_2]^k}{k!} G_0^{(k-1)}(p_2) +$$

$$+O\left(\varepsilon^{\frac{N+3}{3}} \ln^{\pi(N+1)} \frac{1}{\varepsilon}\right) = \sum_{k=1}^{\left]\frac{N+3}{2}\right]} \frac{[E_{N+2}(\varepsilon)-p_2]^k}{k!} G_0^{(k-1)}(p_2) + \mathfrak{S}^*(\omega) +$$

$$+O\left(\varepsilon^{\frac{N+3}{3}} \ln^{\pi(N+1)} \frac{1}{\varepsilon}\right). \tag{6.7}$$

Here only the first term is written in detail; it is obtained by applying the binomial formula; it clearly consists of terms that are independent of ω. It can be verified directly that

$$\sum_{k=1}^{\left]\frac{N+3}{2}\right]} \frac{[E_{N+2}(\varepsilon)-p_2]^k}{k!} G_0^{(k-1)}(p_2) =$$

$$= \sum_{n=2}^{N+2} \varepsilon^{n/3} \sum_{\nu=0}^{\pi(n-2)} \mathcal{B}^*_{n,\nu} \ln^\nu \frac{1}{\varepsilon} + O\left(\varepsilon^{\frac{N+3}{3}} \ln^{\pi(N+1)} \frac{1}{\varepsilon}\right) \tag{6.8}$$

[see (II.20.8)], and the coefficients $\mathcal{B}^*_{n,\nu}$ can be effectively calculated. We have written $\mathfrak{S}^*(\omega)$ for the sum of the terms from the binomial formula, explicitly depending on ω, and of lower order than the remainder term.

Transforming the fourth term in (6.5) in a similar fashion, and putting

$$G_n(y) = \tilde{g}_n(y,\, \hat{x}_0(y),\, \hat{x}_1(y),\, \ldots,\, \hat{x}_n(y)),$$

we obtain

$$\sum_{n=1}^{\left]\frac{N+3}{3}\right]} \varepsilon^n \int_{p_2}^{y} \tilde{g}_n(y,\, \hat{x}_0(y),\, \hat{x}_1(y),\, \ldots,\, \hat{x}_n(y))\, dy =$$

$$= \sum_{n=1}^{\left]\frac{N+3}{3}\right]} \varepsilon^n \sum_{k=1}^{\left]\frac{N+3-3n}{2}\right]} \frac{[E_{N-1}(\varepsilon)-p_2]^k}{k!} G_n^{(k-1)}(p_2) +$$

$$+\mathfrak{S}^{**}(\omega)+O\left(\varepsilon^{\frac{N+3}{3}}\ln^{\pi(N-2)}\frac{1}{\varepsilon}\right)=$$

$$=\sum_{m=5}^{N+2}\varepsilon^{m/3}\sum_{\nu=0}^{\pi(m-2)-1}\mathcal{B}^{**}_{m,\nu}\ln^{\nu}\frac{1}{\varepsilon}+\mathfrak{S}^{**}(\omega)+O\left(\varepsilon^{\frac{N+3}{3}}\ln^{\pi(N-2)}\frac{1}{\varepsilon}\right); \quad (6.9)$$

here the coefficients $\mathcal{B}^{**}_{m,\nu}$ can be effectively calculated. We have written $\mathfrak{S}^{**}(\omega)$ for a sum of terms depending explicitly on ω and of order lower than the remainder.

Lemma 2. The time for the phase point of (1.1) to traverse the drop part $E\hat{A}$ of the trajectory Z_ε lying close to the segment $P^0P\hat{R}^0$ of Z_0 has the asymptotic representation

$$T_{E\hat{A}}=T_{P\hat{R}^0}+\sum_{n=1}^{]\frac{N+3}{3}]}\varepsilon^n\int_{P\hat{R}^0}g_n[f(x,y),g(x,y)]\,dy+$$
$$+\sum_{n=2}^{N+1}\varepsilon^{n/3}\sum_{\nu=0}^{\pi(n-2)}B_{n,\nu}\ln^{\nu}\frac{1}{\varepsilon}+O\left(\varepsilon^{\frac{N+2}{3}}\ln^{\pi(N)}\frac{1}{\varepsilon}\right), \quad (6.10)$$

where N is any positive integer. Integration of (1.1) is not required in the calculation of the coefficients $B_{n,\nu}$.

Since the drop time $T_{E\hat{A}} = T_{EG} = T_{G\hat{A}}$, the lemma can be proved by combining (6.2) and (6.5) and using (6.3), (6.4) and (6.6)-(6.9). It is essential here that all terms depending explicitly on ω vanish, because

$$-\frac{1}{f'_x(P)}\varepsilon\ln\omega+\mathfrak{S}_0(\omega)-\mathfrak{S}^{*}(\omega)-\mathfrak{S}^{**}(\omega)=O\left(\varepsilon^{\frac{N+2}{3}}\ln^{\pi(N)}\frac{1}{\varepsilon}\right). \quad (6.11)$$

This relation can be proved by direct but long calculations, in which the appropriate expressions are written in detail and (II.21.7) and (II.21.8) are used.

The coefficients $B_{n,\nu}$ in (6.10) are calculated directly in terms of values of the right sides in (1.1) and certain of their derivatives on the fast-motion part SP of Z_0. This can be seen by analyzing the determination of the coefficients in (6.3), (6.4), (6.8), and (6.9).

We saw in Chapter II, Sec. 20 that each of the functions $z_{n,\nu}(w)$ can be expressed in terms of values of $f(x,y)$ and $g(x,y)$ and certain of their derivatives on SP. The variable change $x = p_1 + sz_{0,0}(w)$ [see (II.20.11), (II.20.12), and

(II.19.8)] transforms the generalized integral

$$\oint_1^\infty g^*_{n,\nu}(p_1+sz_{6,0}(w),\ z_{2,0}(w),\ \ldots,\ z_{n-5,\varkappa}(w))\,dw$$

[see (6.4)] into a generalized integral, on the segment $P^\circ P$ of the trajectory Z_0, of a function expressed in terms of only $f(x,y)$, $g(x,y)$, and their derivatives. The integral obtained, like the terms outside the integral, in general contains quantities expressed in terms of only these functions and their derivatives on SP°. The constants $\mathcal{B}^*_{n,\nu}$ and $\mathcal{B}^{**}_{n,\nu}$ [see (6.8) and (6.9)] clearly can be expressed in terms of values of $f(x,y)$, $g(x,y)$, and their derivatives on SP [see (II.20.3) and Chapter II, Sec. 17].

An example of the representation (6.10) is the following formula, which is easily proved by using relations (II.20.18), (II.21.6), (II.20.22), (3.4), (II.20.3), and (II.17.15):

$$T_{E\hat{A}}=T_{P\hat{R}^0}+\varepsilon\int\limits_{P\hat{R}^0}\frac{-g'_x(x,y)}{f'_x(x,y)\,g(x,y)}\,dy+B_{2,0}\varepsilon^{2/3}+B_{3,1}\varepsilon\ln\frac{1}{\varepsilon}+B_{3,0}\varepsilon+O(\varepsilon^{4/3}),$$

$$B_{2,0}=-\frac{\gamma^{2/3}\Omega_0}{g(P)}\operatorname{sign}g(S),\qquad B_{3,1}=-\frac{\gamma'_\xi}{3g(P)}\operatorname{sign}g(S),\qquad (6.12)$$

$$B_{3,0}=-\frac{\gamma'_\xi\Omega_1}{g(P)}\operatorname{sign}g(S)+\frac{\gamma'_\xi\ln\gamma}{3g(P)}\operatorname{sign}g(S)-$$

$$-\frac{1}{g(P)}\left[\operatorname{sign}g(S)\oint_0^q\frac{\gamma(\xi,0)}{\xi^2}\,d\xi+\int\limits_{Q+P^0}\frac{g(x,y)}{f(x,y)}\,dx\right]+\oint_1^\infty\frac{b\,dw}{wg(p_1+sz_{0,0}(w),\,p_2)}.$$

The transformation (II.5.3) yields

$$\operatorname{sign}g(S)\oint_0^q\frac{\gamma(\xi,0)}{\xi^2}\,d\xi+\oint\limits_{SQ+}\frac{g(x,y)}{f(x,y)}\,dx+\gamma\frac{\varphi''(0)}{2\varphi'^2(0)}\operatorname{sign}f'_y(S)+\gamma'_\xi\ln\varphi'(0)\operatorname{sign}g(S)$$

(6.13)

(see Chapter II, Sec. 13). Furthermore, by introducing the new variable of integration $x = p_1 + sz_{0,0}(w)$, we obtain

$$\oint_1^\infty\frac{b\,dw}{wg(p_1+sz_{0,0}(w),\,p_2)}=\oint_{p_1^0}^{p_1}\frac{dx}{f(x,p_2)}+\frac{1}{f'_x(P)}\ln z^{0,0}_{1,0}=$$

$$=\oint\limits_{P^0P}\frac{dx}{f(x,y)}-\frac{1}{g(P)}\oint\limits_{P^0P}\frac{g(x,y)}{f(x,y)}\,dx\qquad(6.14)$$

[see (II.19.9), (II.20.11), and (II.21.1)]. Hence the quantities in (6.12) can be found, without integrating the system (1.1), directly in terms of the right sides of the system [see (4.22), (6.13), and (6.14)].

7. An Asymptotic Formula for the Relaxation-Oscillation Period

Here we apply the results obtained in Secs. 3-6 to find an asymptotic formula enabling us to calculate the relaxation-oscillation period described by the nondegenerate system (1.1) with any desired degree of accuracy.

Let Z_0 be a closed trajectory of (1.2) (Fig. 48), and let $S_1,\dots,S_M$ be all the junction points of Z_0, enumerated in the natural order in which they are encountered by the phase point of (1.2). Let $P_1,\dots,P_M$ be the drop points of this trajectory, P_m being the drop point following the junction point S_m, $m = 1,\dots,M$. Then Z_0 can be divided in a natural fashion into M pairs of slow- and fast-motion parts,

$$(P_M S_1;\ S_1 P_1),\ (P_1 S_2;\ S_2 P_2),\ \dots,\ (P_{M-1} S_M;\ S_M P_M). \tag{7.1}$$

Theorem 2. Suppose that the assumptions made in Sec. 1, Chapter II and in Sec. 1 of the present chapter are satisfied. Then the period T_ε of the limit cycle Z_ε of the nondegenerate system (1.1) has the asymptotic representation

$$T_\varepsilon = T_0 + \sum_{m=1}^{M} \Delta_m T + O\left(\varepsilon^{\frac{N+2}{3}} \ln^{\pi(N)} \frac{1}{\varepsilon}\right), \tag{7.2}$$

where N is any positive integer, T_0 is the period of the closed trajectory Z_0 calculated from formula (1.5), and the $\Delta_m T$, $m = 1,\dots,M$ can be expressed in the form

$$\Delta_m T = \sum_{n=2}^{N+1} \varepsilon^{n/3} \sum_{\nu=0}^{\pi(n-2)} \mathcal{K}^{om}_{n,\nu} \ln^{\nu} \frac{1}{\varepsilon}. \tag{7.3}$$

The coefficients $\mathcal{K}^{om}_{n,\nu}$, for each m, $1 \le m \le M$, are expressed directly in terms of values of the right sides in (1.1) and their derivatives on the pair $(P_{m-1}S_m;S_mP_m)$ of adjacent parts of Z_0, so that the determination of these coefficients does not require the integration of the nondegenerate system (1.1).

From the pairs of adjacent parts (7.1) we select any pair $(\tilde{P}S;SP)$, and consider the trajectory Z_ε close to these segments $ABDE\hat{A}$ (see Sec. 2 and Fig. 48). Combining the results (3.3), (4.11), (4.18), (5.2), and (6.10), and taking the accuracy required into account, we easily obtain the following representation for the traversal time of this part by the phase point of (1.1):

$$
\begin{aligned}
T_{A\hat{A}} = T_{R^0S} + \sum_{n=1}^{\left]\frac{N+2}{3}\right[} \varepsilon^n \int_{R^0Q^-} g_n[f(x,y), g(x,y)]\,dy + \\
+ T_{Q^-R} + \sum_{n=1}^{\left]\frac{N+2}{3}\right[} \varepsilon^n \int_{Q^-R} g_n[f(x,y), g(x,y)]\,dy - \\
- \mathfrak{S}_2(-q) + \sum_{n=3}^{N+1} A_n^{(1)}\varepsilon^{n/3} + \sum_{n=1}^{\left]\frac{N+2}{3}\right[} A_n^{(2)}\varepsilon^n \ln\frac{1}{\varepsilon} + \\
+ \sum_{n=2}^{N+1} \varepsilon^{n/3} \sum_{\nu=0}^{\pi(n-2)} A_{n,\nu}^{*} \ln^{\nu}\frac{1}{\varepsilon} + \varepsilon \int_{Q^+P^0} \frac{dx}{f(x,y)} + \\
+ \sum_{n=5}^{N+1} \varepsilon^{n/3} \sum_{\nu=0}^{\pi(n-2)-1} \ln^{\nu}\frac{1}{\varepsilon} \int_{Q^+P^0} f_{n,\nu}[f(x,y), g(x,y)]\,dx + \\
+ \sum_{n=2}^{N+1} \varepsilon^{n/3} \sum_{\nu=0}^{\pi(n-2)} B_{n,\nu} \ln^{\nu}\frac{1}{\varepsilon} + T_{P\hat{R}^0} + \\
+ \sum_{n=1}^{\left]\frac{N+2}{3}\right[} \varepsilon^n \int_{P\hat{R}^0} g_n[f(x,y), g(x,y)]\,dy + O\left(\varepsilon^{\frac{N+2}{3}} \ln^{\pi(N)}\frac{1}{\varepsilon}\right).
\end{aligned}
\tag{7.4}
$$

To find the period T_ε of Z_ε, we must form expressions of the type (7.4) for each of the pairs of adjacent parts in (7.1), and then add all the expressions. It is clear that we can regroup the terms as follows: The last two terms

$$
T_{P\hat{R}^0} + \sum_{n=1}^{\left]\frac{N+2}{3}\right[} \varepsilon^n \int_{P\hat{R}^0} g_n[f(x,y), g(x,y)]\,dy,
$$

written in (7.4), are transferred to the expression corresponding to the pair of adjacent parts following the pair $(\tilde{P}S; SP)$, while the two analogous terms

$$
T_{\tilde{P}R^0} + \sum_{n=1}^{\left]\frac{N+2}{3}\right[} \varepsilon^n \int_{\tilde{P}R^0} g_n[f(x,y), g(x,y)]\,dy \tag{7.5}
$$

from the expression corresponding to the pair of adjacent parts preceding the pair $(\tilde{P}S; SP)$ are added to (7.4). The terms (7.5) are combined respectively with the first and second term of (7.4).

We now consider the third, fourth, and fifth terms in (7.4). Results in Sec. 7 of Chapter II and Sec. 4 of the present chapter imply that, with accuracy of the required order, each of these terms is a polynomial in ε. A more detailed investigation shows that

$$T_{Q-R}+\sum_{n=1}^{\left]\frac{N+2}{3}\right]} \varepsilon^n \int_{Q-R} g_n[f(x, y),\ g(x, x)]\,dy-\mathfrak{S}_2(-q)= \\ =O\left(\varepsilon^{\frac{N+2}{3}} \ln^{\pi(N)}\frac{1}{\varepsilon}\right). \quad (7.6)$$

Hence, by taking the above regrouping into account and using (7.6), we can set up a correspondence between the part $AB DE\hat{A}$ of the trajectory Z_ε and the expression

$$T^*_{A\hat{A}}=T_{\tilde{P}S}+\sum_{n=2}^{N+1}\varepsilon^{n/3}\sum_{\nu=0}^{\pi(n-2)} K_{n,\nu}^{A\hat{A}}\ln^{\nu}\frac{1}{\varepsilon}+O\left(\varepsilon^{\frac{N+2}{3}}\ln^{\pi(N)}\frac{1}{\varepsilon}\right).$$

The coefficients in this expression are clearly determined only by values of $f(x,y)$ and $g(x,y)$ and their derivatives on the part $\tilde{P}SP$ of Z_0. This proves (7.2) and (7.3).

Theorem 2 completely determines the structure of the asymptotic expansion of the period T_ε of relaxation oscillations in a second-order system (1.1) corresponding to a stable limit cycle Z_ε close to the phase trajectory Z_0 of the discontinuous periodic solution of the degenerate system (1.2). This expansion is of the form

$$T_\varepsilon=\sum_{\substack{n=0\\ n\neq 1}}^{\infty}\varepsilon^{n/3}\sum_{\nu=0}^{\pi(n-2)} K_{n,\nu}\ln^{\nu}\frac{1}{\varepsilon}, \quad (7.7)$$

[see (II.12.14)] or, equivalently,

$$\pi(n)=\left[\frac{n}{3}\right]+\begin{cases}0, & \text{if } n\not\equiv 1 \pmod 3,\\ 1, & \text{if } n\equiv 1 \pmod 3,\end{cases}$$

$$\pi(n)=\frac{n}{3}+\frac{2\sqrt{3}}{9}\tan\frac{\pi n}{3}.$$

The series (7.7) is a natural generalization of the usual asymptotic power series and has many properties in common with formal power series.

We stress that the structure (7.7) of the asymptotic expansion for the relaxation-oscillation period was obtained under certain simplifying assumptions (Sec. 1 in Chapter I and Sec. 1 of the present chapter) concerning the system (1.1), in particular the assumption of the nonsingularity of all junction points of Z_0 [see (II.1.6) and (II.1.15)]. The structure of the asymptotic expansion when the irregular points of Z_0 have a more general nature is considered in [51].

The coefficients in (7.7) are determined recursively without the necessity of integrating the nondegenerate system (1.1). If we carry out the more detailed study described in Secs. 3-6, we can find explicit expressions for any required number of these coefficients directly in terms of $f(x,y)$ and $g(x,y)$. For example, if we start from (3.4), (4.21), (5.3), and (6.12), and use (4.22), (6.13), and (6.14), we obtain the following result after some elementary transformations.

Theorem 3. The period T_ε of the limit cycle Z_ε of a nondegenerate system (1.1), satisfying the conditions of Sec. 1, Chapter II and Sec. 1 of the present chapter, has the asymptotic representation

$$T_\varepsilon = \oint_{Z_0} \frac{dy}{g(x, y)} + \sum_{m=1}^{M} \Delta_m T + O(\varepsilon^{4/3}); \tag{7.8}$$

here the $\Delta_m T$, $m = 1,\dots,M$ correspond to the pair $(P_{m-1}S_m; S_mP_m)$ of adjacent parts of Z_0, and are given by the formula

$$\Delta_m T = \mathcal{K}^{m}_{2,0}\varepsilon^{2/3} + \mathcal{K}^{m}_{3,1}\varepsilon \ln\frac{1}{\varepsilon} + \mathcal{K}^{m}_{3,0}\varepsilon, \tag{7.9}$$

where

$$\mathcal{K}^{m}_{2,0} = \gamma^{2/3}(S_m)\,\Omega_0\chi(S_mP_m),$$
$$\mathcal{K}^{m}_{3,1} = \frac{1}{3}\gamma'_\xi(S_m)\,\chi(S_mP_m) + \frac{1}{6}\gamma(S_m)\,\delta'_\xi(S_m),$$

$$\begin{aligned}\mathcal{K}^{m}_{3,0} = {} & \oint_{P_{m-1}S_m} \frac{-g'_x(x, y)}{f'_x(x, y)\,g(x, y)}\,dx + \oint_{S_mP_m} \frac{dx}{f(x, y)} - \frac{1}{g(P_m)} \oint_{S_mP_m} \frac{g(x, y)}{f(x, y)}\,dx + \\ & + \Big[\gamma'_\xi(S_m)\,\Omega_1 - \frac{1}{3}\gamma'_\xi(S_m)\ln\gamma(S_m) + \\ & + \gamma'_\xi(S_m)\ln\varphi'(S_m) + \gamma(S_m)\frac{\varphi''(S_m)}{2\varphi'^2(S_m)}\operatorname{sign} f^2_x(S_m)\Big]\chi(S_mP_m) - \\ & - \gamma(S_m)\,\delta'_\xi(S_m)\Big[I_0 + \frac{1}{6}\ln\gamma(S_m) - \frac{1}{2}\ln\varphi'(S_m)\Big];\end{aligned} \tag{7.10}$$

$$\varphi'(S_m) = \sqrt{\left|\frac{f^{(2)}_x(S_m)}{2f'_y(S_m)}\right|},$$

$$\varphi''(S_m) = \frac{f_x^{(3)}(S_m) f_y'(S_m) - 3f_{xy}''(S_m) f_x^{(2)}(S_m)}{6f_y'^2(S_m)\varphi'(S_m)} \operatorname{sign} g(S_m),$$

$$\chi(S_m P_m) = \left[\frac{1}{g(S_m)} - \frac{1}{g(P_m)}\right] \operatorname{sign} g(S_m),$$

$$\gamma(S_m) = |g(S_m)| \sqrt{\frac{2}{|f_x^{(2)}(S_m) f_y'(S_m)|}},$$

$$\gamma_\xi'(S_m) = \frac{6f_x^{(2)}(S_m) g_x'(S_m) - 2f_x^{(3)}(S_m) g(S_m)}{3[f_x^{(2)}(S_m)]^2} \operatorname{sign} g(S_m),$$

$$\delta_\xi'(S_m) = -\frac{g_x'(S_m)}{g^2(S_m)\varphi'(S_m)} \operatorname{sign} f_y'(S_m),$$

$$\Omega_0 = \lim_{u\to\infty} v_0(u), \quad \boldsymbol{\Omega}_1 = \lim_{u\to\infty} [\boldsymbol{v}_1(u) - \ln u],$$

$$\boldsymbol{I}_0 = 1 + \oint_{-\infty}^{0} z_0(u)\,du + \oint_{0}^{\infty} v_0(u)\,du.$$

8. Van der Pol's Equation. Dorodnitsyn's Formula

We shall apply Theorem 3 to approximate the relaxation-oscillation period of the system described by Van der Pol's equation.

We have already noted in Sec. 3, Chapter I that, for any value of $\lambda > 0$, the Van der Pol equation

$$\frac{d^2x}{dt^2} + \lambda(-1 + x^2)\frac{dx}{dt} + x = 0 \tag{8.1}$$

has a periodic solution [see (I.3.6)]. It is easily seen that, for sufficiently large values of λ, this solution has the nature of a relaxation oscillation.

We introduce another unknown y, a new time t, and a parameter ε defined as follows [cf. (I.3.7)]:

$$y = \int_0^x (x^2 - 1)\,dx + \frac{1}{\lambda}\frac{dx}{dt}, \quad t = \frac{t}{\lambda}, \quad \varepsilon = \frac{1}{\lambda^2}. \tag{8.2}$$

Then (8.1) becomes the second-order system

$$\begin{cases} \varepsilon\dot{x} = y - \frac{1}{3}x^3 + x, \\ \dot{y} = -x \end{cases} \tag{8.3}$$

[cf. (I.3.8)], with a small (positive) parameter multiplying a derivative, i.e., a system of the type investigated in this and the preceding chapter. It can be directly verified that the system (8.3) satisfies all the conditions in Sec. 1 of the present chapter and Sec. 1 of Chapter II. In particular (Chapter I, Sec. 4), the degenerate system

$$\begin{cases} y - \frac{1}{3}x^3 + x = 0, \\ \dot{y} = -x \end{cases} \tag{8.4}$$

has a closed phase trajectory Z_0 (Fig. 8) consisting of two slow-motion parts P_2S_1 and P_1S_2 and two fast-motion parts S_1P_1 and S_2P_2. The coordinates of the junction points S_1 and S_2 and the drop points P_1 and P_2 are as follows:

$$S_1(-1,\ 2/3),\quad S_2(1,\ -2/3),\quad P_1(2,\ 2/3),\quad P_2(-2,\ -2/3). \tag{8.5}$$

Theorem 1 now implies that, for each sufficiently small value of the parameter Z_0, the system (8.3) has a unique and stable limit cycle Z_ε, and $Z_\varepsilon \to Z_0$ uniformly for $\varepsilon \to 0$. In other words, the system described by (8.1) performs stable relaxation oscillations if λ is large enough.

By using the results in Chapter II, we can obtain asymptotic representations, with any required degree of accuracy with respect to ε, for a closed trajectory Z_ε of (8.3), with, for example, accuracy of order $\varepsilon^{4/3}$ for $\varepsilon \to 0$. This solves the problem of the approximate determination of a nearly discontinuous periodic solution of (8.1). In particular, we can find the amplitude of relaxation oscillations. However, we do not dwell on all these problems; we shall investigate only the construction of an asymptotic approximation for the period of relaxation oscillations, with accuracy of order $\varepsilon^{4/3}$.

It follows from (7.8) that the period T_ε of the limit cycle Z_ε of (8.3) has the asymptotic representation

$$T_\varepsilon = T_0 + \Delta_1 T + \Delta_2 T + O(\varepsilon^{4/3}), \tag{8.6}$$

where T_0 is the period of the closed trajectory Z_0 of the system (8.4), and $\Delta_1 T$ and $\Delta_2 T$ correspond to the pairs $(P_2S_1;S_1P_1)$ and $(P_1S_2;S_2P_2)$ of adjacent parts of this trajectory. Clearly

$$T_0 = \oint_{Z_0} \frac{dy}{-x} = 3 - 2\ln 2.$$

Moreover, the phase-velocity field of (8.3) has central symmetry; hence $\Delta_1 T = \Delta_2 T$, so that it is sufficient to calculate $\Delta_1 T$.

By virtue of (8.5), formulas (7.10) imply that

$$\varphi'(S_1)=1,\quad \varphi''(S_1)=-1/3,\quad \chi(S_1P_1)=3/2,$$
$$\gamma(S_1)=1,\quad \gamma'_\xi(S_1)=-2/3,\quad \delta'_\xi(S_1)=1.$$

The values of the generalized integrals appearing in (7.1) are easily calculated (Chapter II, Sec. 13):

$$\oint\limits_{P_2S_1}\frac{dx}{-x(-x^2+1)}=\int\limits_{-2}^{-1}\left(\frac{-1}{x}+\frac{1/2}{x-1}\right)dx-\frac{1}{2}\ln 1=\frac{3}{2}\ln 2-\frac{1}{2}\ln 3,$$

$$\oint\limits_{S_1P_1}\frac{dx}{y-\frac{1}{3}x^3+x}=-3\oint\limits_{-1}^{0}\frac{dx}{(x-2)(x+1)^2}-3\oint\limits_{0}^{2}\frac{dx}{(x-2)(x+1)^2}=\frac{2}{3}\ln 3-\frac{1}{3},$$

$$\oint\limits_{S_1P_1}\frac{-x\,dx}{y-\frac{1}{3}x^3+x}=3\oint\limits_{-1}^{2}\frac{x\,dx}{(x-2)(x+1)^2}=-\frac{4}{3}\ln 3-\frac{1}{3}.$$

We can now write an expression for $\Delta_1 T$ [see (7.9)] and substitute this expression in (8.6).

This yields the following asymptotic representation for the period of a trajectory Z_ε of (8.3):

$$T_\varepsilon=3-2\ln 2+3\Omega_0\varepsilon^{2/3}-\frac{1}{3}\varepsilon\ln\frac{1}{\varepsilon}+$$
$$+\left(3\ln 2-\ln 3-\frac{3}{2}-2\Omega_1-2I_0\right)\varepsilon+O(\varepsilon^{4/3});$$

the universal constants Ω_0, Ω_1 and I_0 here are given in (7.10). We note that the period is calculated with respect to the time t. If we return to the old time t and the parameter λ [see (8.2)], relation (8.7) becomes

$$T_\lambda=(3-2\ln 2)\lambda+3\Omega_0\lambda^{-1/3}-\frac{2}{3}\frac{\ln\lambda}{\lambda}+$$
$$+\left(3\ln 2-\ln 3-\frac{3}{2}-2\Omega_1-2I_0\right)\frac{1}{\lambda}+O(\lambda^{-5/3}).$$

This is Dorodnitsyn's formula for the period of relaxation oscillations of the system described by Van der Pol's equation [18, 58].

CHAPTER IV

SYSTEMS OF ARBITRARY ORDER. ASYMPTOTIC CALCULATION OF SOLUTIONS

In contrast to the case $n = 2$ investigated in the preceding two chapters, the asymptotic representation of trajectories of systems of arbitrary order n with a small parameter multiplying some derivatives has not been studied up to the present. In particular, asymptotic expansions of periodic solutions describing relaxation oscillations have not yet been obtained.

In this chapter we find asymptotic approximations for trajectories of a system of arbitrary order with a small parameter ε multiplying some of the derivatives, with accuracy to within quantities of order ε.

1. Basic Assumptions

We shall study the phase trajectories of the system

$$\begin{cases} \varepsilon \dot{x}^i = f^i(x^1, \ldots, x^k, y^1, \ldots, y^l), \\ \dot{y}^j = g^j(x^1, \ldots, x^k, y^1, \ldots y^l), \end{cases} \tag{1.1}$$

$$i = 1, \ldots, k, \quad j = 1, \ldots, l, \quad k + l = n,$$

where ε is a small positive parameter. We use the notation

$$x = (x^1, \ldots, x^k), \quad f = (f^1, \ldots, f^k),$$

$$y = (y^1, \ldots, y^l), \quad g = (g^1, \ldots, g^l).$$

The right sides in (1.1) are assumed to be defined in the whole phase space R^n of the variables $x^1, \ldots, x^k$, $y^1, \ldots, y^l$ and to be sufficiently smooth, i.e., differentiable with respect to the totality of the variables, the number of times required in our reasoning. The phase space R^n, $n = k + l$, of (1.1) is the direct sum of the k-dimensional subspace X^k and the l-dimensional subspace Y^l.

The nondegenerate system (1.1) corresponds to the *degenerate* system obtained by putting $\varepsilon = 0$ in (1.1):

$$\begin{cases} f^i(x^1, \ldots, x^k, y^1, \ldots, y^l) = 0, & i = 1, \ldots, k, \\ \dot{y}^j = g^j(x^1, \ldots, x^k, y^1, \ldots, y^l), & j = 1, \ldots, l. \end{cases} \quad (1.2)$$

We also use the *fast-motion equation system*

$$\varepsilon \dot{x}^i = f^i(x^1, \ldots, x^k, y^1, \ldots, y^l), \quad i = 1, \ldots, k, \quad (1.3)$$

corresponding to (1.1). This is a normal kth order system in which $y^1, \ldots, y^l$ are assumed to be parameters. The phase of (1.3) is identified with the k-dimensional plane of $X^k_{y^1, \ldots, y^l}$ of R^n formed of the points $(x^1, \ldots, x^k, y^1, \ldots, y^l)$, where $y^1, \ldots, y^l$ are fixed. We assume that (1.3), for arbitrary values of the parameters $y^1, \ldots, y^l$, has only equilibrium positions as stationary solutions.

An important role will be played by the l-dimensional surface Γ determined in R^n by the first k equations of the system (1.2):

$$f^i(x^1, \ldots, x^k, y^1, \ldots, y^l) = 0, \quad i = 1, \ldots, k. \quad (1.4)$$

This surface is clearly the set of all equilibrium positions of (1.3) for all possible values of the parameters $y^1, \ldots, y^l$ (Fig. 24).

Consider the matrix

$$\mathfrak{A}(x^1, \ldots, x^k, y^1, \ldots, y^l) = \left\| \frac{\partial}{\partial x^\beta} f^\alpha(x^1, \ldots, x^k, y^1, \ldots, y^l) \right\|, \quad \alpha, \beta = 1, \ldots, k, \quad (1.5)$$

defined in R^n and, in particular, on Γ. The set of points on the surface (1.4) at which all eigenvalues of the matrix (1.5) have negative real parts will be called the *stable region* of the surface Γ, and will be denoted by Γ_-. We assume henceforth that the stable region Γ_- does not contain any equilibrium position of (1.1).

The points of Γ for which

$$\det \mathfrak{A}(x^1, \ldots, x^k, y^1, \ldots, y^l) = 0 \quad (1.6)$$

will be called *nonregular points*. In general, the set Γ_0 of all nonregular points is an $(l-1)$-dimensional subset of surface Γ and divides it into two or more parts. By virtue of (1.6), at least one eigenvalue of $\mathfrak{A}$ vanishes at each nonregular point.

A nonregular point $S(x_0^1, \ldots, x_0^k, y_0^1, \ldots, y_0^l)$ on the surface Γ will be called a *junction point* if the following conditions are satisfied:

(a) S is an equilibrium position of (1.1), i.e., $g(S) \neq 0$;

(b) all eigenvalues of the matrix $\mathfrak{A}(S)$, except one, which vanishes, have negative real parts;

(c) for $y^1 = y_0^1, \ldots, y^l = y_0^l$, (1.3) has only one trajectory approaching S when $t \to -\infty$;

(d) no k-dimensional plane in R^n obtained by the translation of the plane $X^k_{y_0^1, \ldots, y_0^l}$ by the vector h [an arbitrary vector $g(S)$, sufficiently small but independent of ε] contains equilibrium positions of (1.3) near S in R^n.

Let $(\tilde{x}^1, \ldots, \tilde{x}^k, \tilde{y}^1, \ldots, \tilde{y}^l)$ be a point of R^n not on Γ, and consider the trajectory of (1.3) starting from this given point (in which the parameters $y^1, \ldots, y^l$ are replaced by their values $y^1 = \tilde{y}^1, \ldots, y^l = \tilde{y}^l$). If this trajectory tends, for increasing t, to a stable equilibrium position $(\tilde{\tilde{x}}^1, \ldots, \tilde{\tilde{x}}^k, \tilde{y}^1, \ldots, \tilde{y}^l)$, in the stable region Γ_-, then the point $(\tilde{\tilde{x}}^1, \ldots, \tilde{\tilde{x}}^k, \tilde{y}^1, \ldots, \tilde{y}^l)$ is called the *drop point* corresponding to point $(\tilde{x}^1, \ldots, \tilde{x}^k, \tilde{y}^1, \ldots, \tilde{y}^l)$ (Figs. 14 and 15).

Finally, let $S(x_0^1, \ldots, x_0^k, y_0^1, \ldots, y_0^l)$ be a junction point satisfying the following extra condition: The trajectory of (1.3) with $y^1 = y_0^1, \ldots, y^l = y_0^l$, tending to S for $t \to -\infty$, asymptotically approximates, for $t \to +\infty$, a stable equilibrium position $P(x_*^1, \ldots, x_*^k, y_0^1, \ldots, y_0^l)$ in the stable region Γ_-. Then P will be called the *drop point following the junction point* S (Figs. 14 and 15).

2. The Zeroth Approximation

For the system (1.2), we can introduce the concept of a *discontinuous solution* (Chapter I, Sec. 6); then it is natural to understand phase trajectories of this system to be the trajectories of its discontinuous solutions. In other words, *the phase trajectory of the degenerate system (1.2) with initial point* Q is the continuous curve $\mathfrak{T}$ in R^n obtained by the successive application of the following rules.

(1) If $Q(\tilde{x}^1, \ldots, \tilde{x}^k, \tilde{y}^1, \ldots, \tilde{y}^l)$ is not on Γ, (1.3) has a unique trajectory starting from Q. This trajectory either goes to infinity or approximates some equilibrium position.

We shall consider only the situation in which there is a drop point from Q. Then the trajectory $\mathfrak{T}$, in its first section, has another trajectory of (1.3) in the plane $X^k_{\tilde{y}^1,\ldots,\tilde{y}^l}$, beginning at Q and leading to the drop point from Q. The phase point of (1.2) passes instantaneously through this section of $\mathfrak{T}$ and falls to the stable region Γ_-.

(2) If the point $Q(\tilde{\tilde{x}}^1, \ldots, \tilde{\tilde{x}}^k, \tilde{y}^1, \ldots, \tilde{y}^l)$ is in the stable region Γ_-, then (1.2) has a unique trajectory starting from Q (Chapter I, Sec. 6). This trajectory either always remains in Γ or, at some time, reaches a point $S \in \Gamma_0$. We consider only the case in which S is a junction point. Then the trajectory $\mathfrak{T}$ has, as its first section, an arc QS of a trajectory of (1.2) in the stable region Γ_-. The phase point of (1.2) traverses this section of $\mathfrak{T}$ in a completely determined finite time. Such a section will be called a *slow-motion part* of $\mathfrak{T}$.

(3) If $Q(x_0^1,\ldots,x_0^k,y_0^1,\ldots,y_0^l)$ is a junction point, (1.3) has a unique trajectory approaching Q for $t \to -\infty$. For $t \to +\infty$, this trajectory either goes to infinity or approaches an equilibrium position. We shall consider only the case in which there is a drop point P following the junction point Q. Then the first section of $\mathfrak{T}$ is an arc of a trajectory of (1.3) in the plane $X^k_{y_0^1,\ldots,y_0^l}$, reaching Q and P. The phase point of (1.2) traverses this section of $\mathfrak{T}$ instantaneously and reaches the stable region Γ_-. Such a section will be called a *fast-motion part* of $\mathfrak{T}$.

Figures 14, 15, and 24 show phase trajectories of (1.2) drawn according to the above rules under the foregoing assumptions. We do not determine the phase trajectories of the degenerate system in any other situation.

Now let Q_0 be a fixed point in the phase space R^n, and suppose that the degenerate system (1.2) has a phase trajectory $\mathfrak{T}_0$ starting from this point. Let $\mathfrak{T}_\varepsilon$ be the phase trajectory of the nondegenerate system (1.1) with the same initial point Q_0. The heuristic considerations in Chapter 1, Sec. 5 yield the following result.

Theorem 1. In any bounded part of phase space, the curve $\mathfrak{T}_0$ serves as a zeroth asymptotic approximation for $\mathfrak{T}_\varepsilon$, i.e., $\mathfrak{T}_\varepsilon \to \mathfrak{T}_0$ when $\varepsilon \to 0$, uniformly on any segment of $\mathfrak{T}_0$ of finite length.

The proof of this theorem is not simple. It can be proved relatively easily for a section of $\mathfrak{T}_0$ between two successive junction points, but not reaching these points (which are independent of ε) [57, 12]. For such a section of $\mathfrak{T}_\varepsilon$ we can also obtain a uniform asymptotic approximation with accuracy up to order ε by using a generalization of the algorithm in [13] and [14] for the construction of asymptotic expansions. We shall state the result, but omit the proof because the reasoning required is similar to that applied in [14].

Theorem 2. Let

$$x^i = x^i(t, \varepsilon),\ i = 1, \ldots, k,\ y^j = y^j(t, \varepsilon),\ j = 1, \ldots, l, \tag{2.1}$$

be the solution of (1.1) satisfying the initial conditions

$$\begin{aligned} x^i(t_1, \varepsilon) &= \tilde{x}^i + a^i\varepsilon^{2/3} + b^i\varepsilon \ln(1/\varepsilon) + O(\varepsilon), \quad i = 1, \ldots, k, \\ y^j(t_1, \varepsilon) &= \tilde{y}^j + c^j\varepsilon^{2/3} + d^j\varepsilon \ln(1/\varepsilon) + O(\varepsilon), \quad j = 1, \ldots, l. \end{aligned} \tag{2.2}$$

Let the point $Q(\tilde{x}^1, \ldots, \tilde{x}^k, \tilde{y}^1, \ldots, \tilde{y}^l)$ be at a finite distance from the surface Γ, and let Q correspond to the drop point $P(\tilde{\tilde{x}}^1, \ldots, \tilde{\tilde{x}}^k, \tilde{y}^1, \ldots, \tilde{y}^l)$ in the stable region Γ_-. Also let

$$x^i = \Pi^i x\left(\frac{t - t_1}{\varepsilon}\right), \quad i = 1, \ldots, k,$$

be the solution of the fast-motion equation system (1.3) for $y^1 = \tilde{y}^1, \ldots, y^l = \tilde{y}^l$ with the initial conditions

$$\Pi^i x(0) = \tilde{x}^i - \tilde{\tilde{x}}^i, \quad i = 1, \ldots k.$$

Finally, let

$$x^i = x_0^i(t), \quad i = 1, \ldots, k, \quad y^j = y_0^j(t), \quad j = 1, \ldots, l,$$

be the solution of the degenerate system (1.2) satisfying the initial conditions

$$y_0^j(t_1) = \tilde{y}^j, \quad j = 1, \ldots, l.$$

If the section of the trajectory of (2.3) for $t_1 \leq t \leq t_2$ lies completely in the stable region Γ_- and is outside a finite neighborhood of the set Γ_0, then the solution (2.1) is defined for $t_1 \leq t \leq t_2$ and, uniformly on this interval,

$$\begin{aligned} x^i(t, \varepsilon) &= x_0^i(t) + \Pi^i x\left(\frac{t - t_1}{\varepsilon}\right) + \varphi^i(t, \varepsilon)\varepsilon^{2/3} + \\ &+ \Phi^i(t, \varepsilon)\varepsilon \ln\frac{1}{\varepsilon} + O(\varepsilon), \quad i = 1, \ldots, k, \\ y^j(t, \varepsilon) &= y_0^j(t) + \psi^j(t, \varepsilon)\varepsilon^{2/3} + \Psi^j(t, \varepsilon)\varepsilon \ln\frac{1}{\varepsilon} + O(\varepsilon), \quad j = 1, \ldots, l; \end{aligned}$$

the functions $\varphi^i(t,\varepsilon)$, $\Phi^j(t,\varepsilon)$, $\psi^i(t,\varepsilon)$, and $\Psi^j(t,\varepsilon)$ can be effectively determined. If $\bar{t}_1$ is any number independent of ε and such that $t_1 < \bar{t}_1 < t_2$, then the solution (2.1) has the following representation on the interval $t_1 \le t \le t$:

$$x^i(t,\varepsilon) = x_0^i(t) + O(\varepsilon), \qquad i = 1, \ldots, k,$$
$$y^j(t,\varepsilon) = y_0^j(t) + O(\varepsilon), \qquad j = 1, \ldots, l.$$

This theorem shows that, when it approaches a junction point but is at a small finite distance from the point, the trajectory $\mathfrak{T}_\varepsilon$ differs from $\mathfrak{T}_0$ by a quantity of order ε. The sections of this chapter to follow will be devoted to the determination of the asymptotic approximation of $\mathfrak{T}_\varepsilon$ with quantities of order $\mathfrak{T}_\varepsilon$ neglected, in a finite neighborhood of a junction point. We shall see, in particular, that in such a neighborhood, the corresponding part of $\mathfrak{T}_\varepsilon$ can serve as a zeroth approximation for $\mathfrak{T}_0$. We shall also prove that, at a small finite distance from a junction point, $\mathfrak{T}_\varepsilon$ has a representation of the type (2.2); hence we can apply Theorem 2 to calculate the next part of $\mathfrak{T}_\varepsilon$.

3. Local Coordinates in the Neighborhood of a Junction Point

For the asymptotic determination of trajectories $\mathfrak{T}_\varepsilon$ of (1.1) near a junction point, we replace, in a small but finite (i.e., independent of ε) neighborhood of this point, the coordinates $x^1,\ldots,x^k$, $y^1,\ldots,y^l$ by local coordinates $\xi^1,\ldots,\xi^k$, $\eta^1,\ldots,\eta^l$, in which (1.1) can be written in a rather simple form.

Let $S(x_0^1,\ldots,x_0^k,y_0^1,\ldots,y_0^l)$ be a junction point, and expand the right sides in (1.1) in a Taylor series in the neighborhood of S; we write only the terms needed in the sequel:

$$f^i(x^1, \ldots, x^k, y^1, \ldots, y^l) =$$
$$= A^i_\alpha(x^\alpha - x_0^\alpha) + B^i_\varkappa(y^\varkappa - y_0^\varkappa) + A^i_{\alpha\beta}(x^\alpha - x_0^\alpha)(x^\beta - x_0^\beta) +$$
$$+ A^i_{\alpha\beta\gamma}(x^\alpha - x_0^\alpha)(x^\beta - x_0^\beta)(x^\gamma - x_0^\gamma) + \ldots, \qquad i = 1, \ldots, k, \tag{3.1}$$
$$g^j(x^1, \ldots, x^k, y^1, \ldots, y^l) = g^j(x_0^1, \ldots, x_0^k, y_0^1, \ldots, y_0^l) + C^j_\alpha(x^\alpha - x_0^\alpha) + \ldots,$$
$$j = 1, \ldots, l.$$

Here we use tensor notation and the summation convention concerning repeated indices; for example,

$$A^i_\alpha(x^\alpha - x_0^\alpha) = \sum_{\alpha=1}^{k} A^i_\alpha(x^\alpha - x_0^\alpha);$$
$$B^i_\varkappa(y^\varkappa - y_0^\varkappa) = \sum_{\varkappa=1}^{l} B^i_\varkappa(y^\varkappa - y_0^\varkappa).$$

The definition of a junction point S implies that the matrix

$$\mathfrak{A}(S) = \| A_\alpha^i \| = \left\| \frac{\partial f^i (x_0^1, \ldots, x_0^k, y_0^1, \ldots, y_0^l)}{\partial x^\alpha} \right\| \tag{3.2}$$

has only one zero eigenvalue (or multiplicity 1). Let

$$m = (m^1, \ldots, m^k)$$

be the corresponding eigenvector. The eigenvector corresponding to the zero eigenvalue of the transposed matrix will be denoted by

$$n = (n_1, \ldots, n_k).$$

The vector n is uniquely determined by the normalization

$$m^\alpha n_\alpha = 1; \tag{3.3}$$

hence

$$A_\alpha^i m^\alpha = 0, \quad A_i^\alpha n_\alpha = 0, \quad i = 1, \ldots, k. \tag{3.4}$$

Consider a linearly independent vector system

$$e_1, \ldots, e_k; \; e_i = (e_i^1, \ldots, e_i^k), \; i = 1, \ldots, k, \tag{3.5}$$

where $e_1 = m$, and the other e_i are such that $(e_i, n) = 0$ but are otherwise arbitrary. Also consider a linearly independent system of vectors

$$h_1, \ldots, h_l; \; h_j = (h_j^1, \ldots, h_j^l), \; j = 1, \ldots, l, \tag{3.6}$$

where $h_1 = g = (g^1, \ldots, g^l) = g(x_0^1, \ldots, x_0^k, y_0^1, \ldots, y_0^l)$, and the other h_j are arbitrary. Let the expansion of $x - x_0$ in the basis (3.5) have the coefficients $\bar{\xi}^1, \ldots, \bar{\xi}^k$, and let the expansion of $y - y_0$ in the basis (3.6) have the coefficients $\bar{\eta}^1, \ldots, \bar{\eta}^l$:

$$\begin{aligned} x^i - x_0^i &= \bar{\xi}^\alpha e_\alpha^i, \quad i = 1, \ldots, k; \\ y^j - y_0^j &= \bar{\eta}^\beta h_\beta^j, \quad j = 1, \ldots, l. \end{aligned} \tag{3.7}$$

Using (3.1), (3.3), and (3.4), and applying obvious transformations, we find that, in the coordinates $\bar{\xi}^1, \ldots, \bar{\xi}^k$, $\bar{\eta}^1, \ldots, \bar{\eta}^l$, (1.1) becomes

$$\begin{cases} \varepsilon \dot{\bar{\xi}}^1 = p(\bar{\xi}^1)^2 + q\bar{\eta}^1 + \bar{b}_{\varkappa'}^1 \bar{\eta}^{\varkappa'} + \bar{d}_1^1 (\bar{\xi}^1)^3 + \bar{c}_\varkappa^1 \bar{\xi}^1 \bar{\eta}^\varkappa + \bar{r}_{\alpha'}^1 \bar{\xi}^1 \bar{\xi}^{\alpha'} + \ldots, \\ \varepsilon \dot{\bar{\xi}}^i = \bar{a}_{\alpha'}^i \bar{\xi}^{\alpha'} + \bar{b}_\varkappa^i \bar{\eta}^\varkappa + \bar{c}_0^i (\bar{\xi}^1)^2 + \bar{d}_1^i (\bar{\xi}^1)^3 + \bar{c}_\varkappa^i \bar{\xi}^1 \bar{\eta}^\varkappa + \bar{r}_{\alpha'}^i \bar{\xi}^1 \bar{\xi}^{\alpha'} + \ldots, \quad i = 2, \ldots, k, \\ \dot{\bar{\eta}}^1 = 1 + \bar{g}_1^1 \bar{\xi}^1 + \ldots, \\ \dot{\bar{\eta}}^j = \bar{g}_1^j \bar{\xi}^1 + \ldots, \quad j = 2, \ldots, l. \end{cases} \tag{3.8}$$

It is easily seen that here

$$\begin{aligned} &p = n_\gamma A^\gamma_{\alpha\beta} m_\alpha m^\beta, \quad q = n_\alpha B^\alpha_\varkappa g^\varkappa, \quad \bar{d}^1_1 = n_\delta A^\delta_{\alpha\beta\gamma} m^\alpha m^\beta m^\gamma, \\ &\bar{b}^1_j = n_\alpha B^\alpha_\varkappa h^\varkappa_j, \qquad j = 2, \ldots, l, \\ &\bar{r}^1_i = 2 n_\gamma A^\gamma_{\alpha\beta} m^\alpha e^\beta_i, \quad i = 2, \ldots, k, \end{aligned} \tag{3.9}$$

and the $\bar{g}^j_1$, $j = 1,\ldots,l$, are the coefficients of the expansion of the vector

$$\hbar = \| C^j_\alpha \| m, \quad \hbar = (\hbar^1, \ldots, \hbar^l), \quad \hbar^j = C^j_\alpha m^\alpha,$$

in the basis $h_1,\ldots,h_l$. In (3.8) and everywhere in the sequel, summation over values of a primed index starts with the value 2; for example,

$$\bar{a}^i_{\alpha'} \bar{\xi}^{\alpha'} = \sum_{\alpha'=2}^{k} \bar{a}^i_{\alpha'} \bar{\xi}^{\alpha'}; \quad \bar{b}^1_{\varkappa'} \bar{\eta}^{\varkappa'} = \sum_{\varkappa'=2}^{l} \bar{b}^1_{\varkappa'} \bar{\eta}^{\varkappa'}.$$

In the first of Eqs. (3.8), the coefficients of $\bar{\xi}^1,\ldots,\bar{\xi}^k$ vanish;

$$\left\| \begin{matrix} 0 & 0 & \ldots & 0 \\ 0 & & & \\ \cdot & & & \\ \cdot & & \| \bar{a}^{i'}_{\alpha'} \| & \\ \cdot & & & \\ 0 & & & \end{matrix} \right\|_{\substack{i'=2,\ldots,k \\ \alpha'=2,\ldots,k}}$$

$\| \bar{a}^{i'}_{\alpha'} \|$ have negative real parts.

We assume below that the junction point S satisfies the following *nondegeneracy condition*:

$$p \neq 0, \quad q \neq 0.$$

The system (3.8) can be simplified further. If we put

$$\begin{aligned} &\bar{\xi}^1 = p^{-2/3} q^{1/3} \xi^1, \qquad \bar{\xi}^i = \xi^i, \quad i = 2, \ldots, k, \\ &\bar{\eta}^1 = p^{-1/3} q^{-1/3} \eta^1, \quad \bar{\eta}^j = \eta^j, \quad j = 2, \ldots, l, \\ &\bar{t} = p^{1/3} q^{1/3} t, \end{aligned} \tag{3.10}$$

(3.8) can be written as

$$\left\{ \begin{aligned} &\varepsilon \dot{\xi}^1 = (\xi^1)^2 + \eta^1 + b^1_{\varkappa'} \eta^{\varkappa'} + d^1_1 (\xi^1)^3 + c^1_\varkappa \xi^1 \eta^\varkappa + r^1_{\alpha'} \xi^1 \xi^{\alpha'} + \ldots, \\ &\varepsilon \dot{\xi}^i = a^i_{\alpha'} \xi^{\alpha'} + b^i_\varkappa \eta^\varkappa + c^i_0 (\xi^1)^2 + d^i_1 (\xi^1)^3 + \\ &\qquad + c^i_\varkappa \xi^1 \eta^\varkappa + r^i_{\alpha'} \xi^1 \xi^{\alpha'} + \ldots, \quad i = 2, \ldots, k, \\ &\dot{\eta}^1 = 1 + g^1_1 \xi^1 + \ldots, \\ &\dot{\eta}^j = g^j_1 \xi^1 + \ldots, \qquad j = 2, \ldots, l. \end{aligned} \right. \tag{3.11}$$

Here the dot indicates differentiation with respect to $\bar{t}$, and

$$\begin{aligned} &b_j^1 = \bar{b}_j^1 p^{1/3} q^{-2/3}, \quad g_1^j = \bar{g}_1^j p^{-1}, \quad j = 2, \ldots, l, \\ &a_{\alpha'}^i = \bar{a}_{\alpha'}^i p^{-1/3} q^{-1/3}, \quad c_0^i = \bar{c}_0^i p^{-5/3} q^{1/3}, \quad i = 2, \ldots, k, \\ &d_1^1 = \bar{d}_1^1 p^{-5/3} q^{1/3}, \quad r_{\alpha'}^1 = \bar{r}_{\alpha'}^1 p^{-1/3} q^{-1/3}, \\ &g_1^1 = \bar{g}_1^1 p^{-2/3} q^{1/3}. \end{aligned} \tag{3.12}$$

We call (3.11) the *canonical form* of (1.1) in the neighborhood of a junction point. Although dots indicate differentiation with respect to $\bar{t}$ in (3.11), we shall simplify our notation in the following, replacing $\bar{t}$ everywhere by t; we return to the actual time t [see (3.10)] only in Chapter V. For the sake of brevity we denote the right sides in (3.11) by Φ^i, Ψ^j:

$$\begin{aligned} &\Phi^1 \equiv (\xi^1)^2 + \eta^1 + b_{\varkappa'}^1 \eta^{\varkappa'} + c_\varkappa^1 \xi^1 \eta^\varkappa + d_1^1 (\xi^1)^3 + r_{\alpha'}^1 \xi^1 \xi^{\alpha'} + \ldots, \\ &\Phi^i \equiv a_{\alpha'}^i \xi^{\alpha'} + b_\varkappa^i \eta^\varkappa + c_0^{il} (\xi^1)^2 + \ldots, \quad i = 2, \ldots, k, \\ &\Psi^1 \equiv 1 + g_1^1 \xi^1 + \ldots, \\ &\Psi^j \equiv g_1^j \xi^1 + \ldots, \quad j = 2, \ldots, l. \end{aligned} \tag{3.13}$$

4. Asymptotic Approximations of a Trajectory at the Beginning of a Junction Section

In a small finite neighborhood of the junction point S, we transform (1.1) into

$$\begin{cases} \varepsilon \dot{\xi}^i = \Phi^i (\xi^1, \ldots, \xi^k, \eta^1, \ldots, \eta^l), & i = 1, \ldots, k, \\ \dot{\eta}^j = \Psi^j (\xi^1, \ldots, \xi^k, \eta^1, \ldots, \eta^l), & j = 1, \ldots, l; \end{cases} \tag{4.1}$$

see (3.13); the junction point is now at the origin. The segment of $\mathfrak{T}_\varepsilon$, in this neighborhood will be called the *junction part.*

In conjunction with (4.1), we consider the corresponding degenerate system

$$\begin{cases} \Phi^i (\xi^1, \ldots, \xi^k, \eta^1, \ldots, \eta^l) = 0, & i = 1, \ldots, k, \\ \dot{\eta}^j = \Psi^j (\xi^1, \ldots, \xi^k, \eta^1, \ldots, \eta^l), & j = 1, \ldots, l. \end{cases} \tag{4.2}$$

Let

$$\xi^i = \xi_0^i (t), \quad i = 1, \ldots, k, \quad \eta^j = \eta_0^j (t), \quad j = 1, \ldots, l, \tag{4.3}$$

be the solution of this degenerate system defined for $-t_1 \leq t < 0$ and satisfying the end conditions

$$\xi_0^i(0)=0, \quad i=1, \ldots, k, \quad \eta_0^j(0)=0, \quad j=1, \ldots, l;$$

the functions (4.3) describe the part of $\mathfrak{T}_0$ adjacent to the junction point in the new coordinates. We first calculate the trajectory of the solution (4.3); we use the independent variable ξ^1.

It is clear that, for $\xi^1 = \ldots = \xi^k = \eta^1 = \ldots = \eta^l = 0$, the Jacobian

$$\begin{vmatrix} \frac{\partial\Phi^1}{\partial\eta^1} & \frac{\partial\Phi^1}{\partial\xi^2} & \cdots & \frac{\partial\Phi^1}{\partial\xi^k} \\ \frac{\partial\Phi^2}{\partial\eta^1} & \frac{\partial\Phi^2}{\partial\xi^2} & \cdots & \frac{\partial\Phi^2}{\partial\xi^k} \\ \cdot & \cdot & \cdot & \cdot \\ \frac{\partial\Phi^k}{\partial\eta^1} & \frac{\partial\Phi^k}{\partial\xi^2} & \cdots & \frac{\partial\Phi^k}{\partial\xi^k} \end{vmatrix}$$

does not vanish. Hence the first k relations (4.2) can be used to express $\eta^1, \xi^2, \ldots, \xi^k$ in terms of $\xi^1, \eta^2, \ldots, \eta^l$. It is easily verified that the resulting expressions are of the form

$$\begin{aligned} \eta^1 &= -(\xi^1)^2 + h^1(\xi^1, \eta^2, \ldots, \eta^l), \\ \xi^i &= k^i(\xi^1)^2 + h^i(\xi^1, \eta^2, \ldots, \eta^l), \quad i=2, \ldots, k; \end{aligned} \tag{4.4}$$

here the k^i are numerical coefficients and the functions $h^1, \ldots, h^k$ do not contain terms $a\xi^1$ or $b(\xi^1)^2$, where a and b are constants.

Differentiation of the first of relations (4.4) with respect to t yields

$$\dot{\eta}^1 = -2\xi^1\dot{\xi}^1 + h^1_{\xi^1}\dot{\xi}^1 + h^1_{\eta^{\varkappa'}}\dot{\eta}^{\varkappa'};$$

hence

$$\dot{\xi}^1 = \frac{\dot{\eta}^1 - h^1_{\eta^{\varkappa'}}\dot{\eta}^{\varkappa'}}{-2\xi^1 + h^1_{\xi^1}},$$

or, if $\dot{\eta}^1$ and $\dot{\eta}^{\kappa'}$ are replaced by their values given in (4.1) and (3.13),

$$\dot{\xi}^1 = \frac{1 + g^1_1\xi^1 + \ldots - h^1_{\eta^{\varkappa'}}(g^{\varkappa'}_1\xi^1 + \ldots)}{-2\xi^1 + h^1_{\xi^1}}. \tag{4.5}$$

Moreover, by using (4.4) for $\eta^1, \xi^2, \ldots, \xi^k$ in (4.5) and in the last l equations (4.2), we obtain the following system for $\xi^1, \eta^2, \ldots, \eta^l$:

$$\begin{cases} \dot{\xi}^1 = \dfrac{1 + g_1^1 \xi^1 + \ldots}{-2\xi^1 + \ldots}, \\ \dot{\eta}^j = g_1^j \xi^1 + \ldots, \quad j = 2, \ldots, l; \end{cases} \tag{4.6}$$

here the dots denote terms not containing $a\xi^1$ or $b\xi^1$, where a and b are constants.

In a sufficiently small finite neighborhood of the point $\xi^1 = \eta^2 = \ldots = \eta^l = 0$, the right side in the first equation in (4.6) is positive for $\xi^1 < 0$; hence, for $-\rho \leq \xi^1 \leq 0$, where ρ is sufficiently small but independent of ε, we can use ξ^1 as the independent variable and replace (4.6) by the system

$$\frac{d\eta^j}{d\xi^1} = \frac{-2g_1^j (\xi^1)^2 + \ldots}{1 + g_1^1 \xi^1 + \ldots}, \quad j = 2, \ldots, l. \tag{4.7}$$

The solution of (4.7) passing through $\eta^2 = \ldots = \eta^l = 0$ for $\xi^1 = 0$ is

$$\eta_0^j (\xi^1) = -\frac{2}{3} g_1^j (\xi^1)^3 + \ldots, \quad j = 2, \ldots, l; \tag{4.8}$$

here the dots denote higher powers of ξ^1. The use of (4.8) on the right-hand sides of (4.4) yields expansions for the trajectory of the solution of (4.3) in powers of ξ^1. We write these expansions with third-degree terms,

$$\begin{aligned} \xi_0^i (\xi^1) &= k^i (\xi^1)^2 + l^i (\xi^1)^3 + \ldots, & i &= 2, \ldots, k, \\ \eta_0^1 (\xi^1) &= -(\xi^1)^2 + \left(\frac{2}{3} b_{\varkappa'}^1 g_1^{\varkappa'} + c_1^1 - d_1^1 - r_{\alpha'}^1 k^{\alpha'} \right) (\xi^1)^3 + \ldots, \\ \eta_0^j (\xi^1) &= -\frac{2}{3} g_1^j (\xi^1)^3 + \ldots, & j &= 2, \ldots, l; \end{aligned} \tag{4.9}$$

here the l^i are constants.

This completes the calculation of the trajectory of the solution (4.3); it is determined for $-\rho \leq \xi^1 \leq 0$.

We now return to the system (4.1). In a sufficiently small but finite neighborhood U_0 of the origin, we use ξ^1 as the independent variable, and, instead of (4.1), we have the system

$$\begin{cases} \dfrac{d\xi^i}{d\xi^1} = \dfrac{\Phi^i (\xi^1, \ldots, \xi^k, \eta^1, \ldots, \eta^l)}{\Phi^1 (\xi^1, \ldots, \xi^k, \eta^1, \ldots, \eta^l)}, & i = 2, \ldots, k, \\ \dfrac{d\eta^j}{d\xi^1} = \varepsilon \dfrac{\Psi^j (\xi^1, \ldots, \xi^k, \eta^1, \ldots, \eta^l)}{\Phi^1 (\xi^1, \ldots, \xi^k, \eta^1, \ldots, \eta^l)}, & j = 1, \ldots, l. \end{cases} \tag{4.10}$$

Let

$$\begin{aligned} \xi^i &= \xi^i(\xi^1, \varepsilon), \quad i=2, \ldots, k, \\ \eta^j &= \eta^j(\xi^1, \varepsilon), \quad j=1, \ldots, l, \end{aligned} \tag{4.11}$$

be a solution of (4.10) with initial point (for $\xi^1 = -\rho$) in U_0 and at a distance from the initial point (for $\xi^1 = -\rho$) of the trajectory (4.9) of order ε. It is natural to try to represent this solution approximately by the sums

$$\begin{aligned} \xi^{i,N} &= \xi_0^i(\xi^1) + \varepsilon\xi_1^i(\xi^1) + \ldots + \varepsilon^N\xi_N^i(\xi^1), \quad i=2, \ldots, k, \\ \eta^{j,N} &= \eta_0^j(\xi^1) + \varepsilon\eta_1^j(\xi^1) + \ldots + \varepsilon^N\eta_N^j(\xi^1), \quad j=1, \ldots, l, \end{aligned} \tag{4.12}$$

where the functions $\xi_\nu^i(\xi^1)$, $\eta_\nu^j(\xi)$, $i = 2,\ldots,k$, $j = 1,\ldots,l$, $\nu = 0, 1,\ldots,N$, are obtained by inserting (4.12) into (4.10) and equating coefficients of powers of ε.

However the functions $\xi_\nu^i(\xi^1)$ and $\eta_\nu^j(\xi^1)$ have power and logarithmic singularities for $\xi^1 = 0$, whose order increases with increasing ν. Hence we must abandon the approximation of the solution (4.11) by sums of the form (4.12) on the whole interval $-\rho \leq \xi^1 \leq 0$. However, we can proceed as follows (Chapter II, Secs. 6 and 7): We consider the quantity $\sigma_1 = \varepsilon^{\lambda_1}$, and select $\lambda_1 > 0$ so that, for $-\rho \leq \xi^1 \leq -\sigma_1$, each successive approximation differs uniformly from the preceding approximation by a quantity of higher order than this preceding approximation. Since we wish to calculate the trajectory $\mathfrak{T}_\varepsilon$ with an error of order at most ε, we conclude from further calculations that it is sufficient to put $\sigma_1 = \varepsilon^{2/7}$, and find only the first and second approximations, i.e., to calculate the sums

$$\begin{aligned} \xi^{i,2} &= \xi_0^i(\xi^1) + \varepsilon\xi_1^i(\xi^1) + \varepsilon^2\xi_2^i(\xi^1), \quad i=2, \ldots, k, \\ \eta^{j,2} &= \eta_0^j(\xi^1) + \varepsilon\eta_1^j(\xi^1) + \varepsilon^2\eta_2^j(\xi^1), \quad j=1, \ldots, l. \end{aligned} \tag{4.13}$$

The functions $\xi_0^i(\xi^1)$ and $\eta_0^j(\xi^1)$ have already been determined by solving the degenerate system (4.2), and are given by (4.9). We shall calculate the functions $\xi_1^i(\xi^1)$, $i = 2,\ldots, k$, and $\eta_1^j(\xi^1)$, $j = 1,\ldots,l$.

Using (4.13) in both sides of the equation

$$\frac{d\eta^1}{d\xi^1} = \varepsilon\,\frac{\Psi^1(\xi^\alpha, \eta^\varkappa)}{\Phi^1(\xi^\alpha, \eta^\varkappa)}$$

and equating the free terms in the expansions in powers of

ε, we find that

$$\frac{d\eta_0^1(\xi^1)}{d\xi^1}=\frac{\Psi^1(\xi^1,\xi_0^{\alpha'}(\xi^1),\eta_0^{\varkappa}(\xi^1))}{\delta\Phi^1},$$

where

$$\delta\Phi^1=\frac{\partial}{\partial\xi^{\beta'}}[\Phi^1(\xi^1,\xi_0^{\alpha'}(\xi^1),\ \eta_0^{\varkappa}(\xi^1))]\,\xi_1^{\beta'}+$$
$$+\frac{\partial}{\partial\eta^{\vartheta}}[\Phi^1(\xi^1,\xi_0^{\alpha'}(\xi^1),\ \eta_0^{\varkappa}(\xi^1)]\,\eta_1^{\vartheta}; \qquad (4.14)$$

hence

$$\delta\Phi^1=\frac{1}{(\eta_0^1)'}\Psi^1(\xi^1,\xi_0^{\alpha'}(\xi^1),\ \eta_0^{\varkappa}(\xi^1)). \qquad (4.15)$$

In these formulas we have used an abbreviated notation for the arguments of the functions Φ^1 and Ψ^1: Instead of $(\xi^1,\ldots,\xi^k,\eta^1,\ldots,\eta^l)$ we have written (ξ^α,η^κ), and instead of $(\xi^2,\ldots,\xi^k)$ we have written $(\xi^{\alpha'})$. We shall use the same notation below.

Moreover, using (4.13) in both sides of the last $l-1$ equations of (4.10), equating the free terms and coefficients of ε in the expansion in powers of ε, and carrying out some elementary transformations, we find that

$$\frac{d\eta_0^j(\xi^1)}{d\xi^1}=\frac{\Psi^j(\xi^1,\xi_0^{\alpha'}(\xi^1),\eta_0^{\varkappa}(\xi^1))}{\delta\Phi^1},$$
$$\frac{d\eta_1^j(\xi^1)}{d\xi^1}=\frac{(\eta_0^1)'}{\Psi^1(\xi^1,\xi_0^{\alpha'}(\xi^1),\eta_0^{\varkappa}(\xi^1))}[\delta\Psi^j-(\eta_0^j)'\,\delta^2\Phi^1]. \qquad (4.16)$$

Here $\delta\Psi^j$ denotes the coefficient of ε in the expansion of $\Psi^j(\xi^1,\xi^{\alpha',2},\eta^{\kappa,2})$ and $\delta^2\Phi^1$ denotes the coefficient of ε^2 in the expansion of $\Phi^1(\xi^1,\xi^{\alpha',2},\eta^{\kappa,2})$. Since

$$(\eta_0^1)'+\varepsilon(\eta_1^1)'=\frac{\Psi^1(\xi^1,\xi_0^{\alpha'},\eta_0^{\varkappa})+\varepsilon\delta\Psi^1+\ldots}{\delta\Phi^1+\varepsilon\delta^2\Phi^1+\ldots},$$

we have

$$\delta\Psi^1=(\eta_0^1)'\,\delta^2\Phi^1+(\eta_1^1)'\,\delta\Phi^1;$$

hence

$$\delta^2\Phi^1=\frac{\delta\Psi^1-(\eta_1^1)'\,\delta\Phi^1}{(\eta_0^1)'}. \qquad (4.17)$$

Using (4.17) in (4.16) and carrying out some elementary calculations, we obtain

$$(\eta_1^j)' = \frac{1}{\Psi^1}[(\eta_0^1)'\,\delta\Psi^j - (\eta_0^j)'\,\delta\Psi^1] + \frac{(\eta_0^j)'\,(\eta_1^1)'}{(\eta_0^1)'}, \quad j=2, \ldots, l. \quad (4.18)$$

Similarly, the first $k-1$ equations in (4.10) yield $(\xi_0^i)' = \delta\Phi^i/\delta\Phi^1$, hence, by virtue of (4.15)

$$\frac{\partial\Phi^i}{\partial\xi^{\beta'}}\xi_1^{\beta'} + \frac{\partial\Phi^i}{\partial\eta^\vartheta}\eta_1^\vartheta = (\xi_0^i)'\,\frac{\Psi^1(\xi^1, \xi_0^{\alpha'}, \eta_0^\varkappa)}{(\eta_0^1)'}, \quad i=2, \ldots, k. \quad (4.19)$$

Moreover, for $i = 1$ [see (4.15) and (4.14)],

$$\frac{\partial\Phi^1}{\partial\xi^{\beta'}}\xi_1^{\beta'} + \frac{\partial\Phi^1}{\partial\eta^1}\eta_1^1 + \frac{\partial\Phi^1}{\partial\eta^{\vartheta'}}\eta_1^{\vartheta'} = \frac{\Psi^1(\xi^1, \xi_0^{\alpha'}, \eta_0^\varkappa)}{(\eta_0^1)'}. \quad (4.20)$$

We now put

$$\frac{\partial\Phi^1}{\partial\eta^1} = 1 + B^1(\xi^1), \qquad \frac{\partial\Phi^1}{\partial\xi^{\beta'}} = D_{\beta'}^1(\xi^1);$$

clearly $B^1(0) = 0$ and $D_{\beta'}^1(0) = 0$. Combining (4.19), (4.20), and (4.18), we arrive at the following equation system for the required functions $\xi_1^2(\xi^1), \ldots, \xi_1^k(\xi^1), \eta_1^1(\xi^1), \ldots, \eta_1^l(\xi^1)$:

$$\begin{cases} \dfrac{\partial\Phi^i}{\partial\xi^{\beta'}}\xi_1^{\beta'} + \dfrac{\partial\Phi^i}{\partial\eta^\vartheta}\eta_1^\vartheta = (\xi_0^i)'\,\dfrac{\Psi^1}{(\eta_0^1)'}, \quad i=2, \ldots, k, \\ (1 + B^1(\xi^1))\,\eta_1^1 = -\dfrac{\partial\Phi^1}{\partial\eta^{\vartheta'}}\eta_1^{\vartheta'} + D_{\beta'}^1(\xi^1)\,\xi_1^{\beta'} + \dfrac{\Psi^1}{(\eta_0^1)'}, \\ (\eta_1^j)' = \dfrac{1}{\Psi^1}[(\eta_0^1)'\delta\Psi^j - (\eta_0^j)'\delta\Psi^1] + \dfrac{(\eta_0^j)'\,(\eta_1^1)'}{(\eta_0^1)'}, \quad j=2, \ldots, l. \end{cases} \quad (4.21)$$

The first k relations of this system are algebraic equations, while the last $l-1$ relations are differential equations. Using the first $k-1$ equations in (4.21) to express the functions $\xi_1^{\beta'}$, $\beta' = 2, \ldots, k$, in terms of η_1^κ, $\kappa = 1, \ldots, l$, and substituting the resulting expressions

$$\xi_1^{\beta'} = G_\varkappa^{\beta'}(\xi^1)\,\eta_1^\varkappa + E^{\beta'}(\xi^1), \quad \beta' = 2, \ldots, k, \quad (4.22)$$

in the right sides of the last $l-1$ equations in (4.21), we obtain, after some simple calculations, the relation

$$(\eta_1^j)' = \tilde{N}^j(\xi^1) + \tilde{N}_\vartheta^j(\xi^1)\,\eta_1^\vartheta + \frac{(\eta_0^j)'\,(\eta_1^1)'}{(\eta_0^1)'}, \quad j=2, \ldots, l; \quad (4.23)$$

it is now easily verified that

$$\tilde{N}^j(0) = \tilde{N}^j_\vartheta(0) = 0, \quad j = 2, \ldots, l, \quad \vartheta = 1, \ldots, l.$$

Furthermore, using (4.22) in the kth equation in (4.21), differentiating the resulting relation with respect to ξ^1, and employing the expression thus obtained for $(\eta^1_1)'$ in (4.23), we obtain the following system of equations for $\eta^2_1, \ldots, \eta^l_1$:

$$(\eta^j_1)' = P^j(\xi^1) + N^j_{\vartheta'}(\xi^1)\,\eta^{\vartheta'}_1 + Q^j_{\vartheta'}(\xi^1)\,(\eta^{\vartheta'}_1)', \quad j = 2, \ldots, l. \tag{4.24}$$

It is easily seen that the functions

$$Q^j_{\vartheta'}(\xi^1), \quad N^j_{\vartheta'}(\xi'), \quad j = 2, \ldots, l, \quad \vartheta' = 2, \ldots, l,$$

vanish for $\xi^1 = 0$, and the functions $P^j(\xi^1)$, $j = 2, \ldots, l$, have second-order poles for $\xi^1 = 0$ with principal parts $g^j_1/2\xi^1$ respectively.

The existence of poles of the right sides of the system (4.24) for $\xi^1 = 0$ means that we can seek a solution of this system in the form

$$\eta^j_1 = K^j(\xi^1)\ln|\xi^1| + L^j(\xi^1), \quad j = 2, \ldots, l. \tag{4.25}$$

Employment of (4.25) in (4.24) yields

$$(K^j)'\ln|\xi^1| + \frac{K^j}{\xi^1} + (L^j)' = P^j + N^j_{\vartheta'}K^{\vartheta'}\ln|\xi^1| +$$
$$+ N^j_{\vartheta'}L^{\vartheta'} + Q^j_{\vartheta'}\left[(K^{\vartheta'})'\ln|\xi^1| + \frac{K^{\vartheta'}}{\xi^1} + (L^{\vartheta'})'\right].$$

Comparison of coefficients of $\ln|\xi^1|$ leads to a system of equations for the functions $K^j(\xi^1)$

$$(K^j)' = N^j_{\vartheta'}K^{\vartheta'} + Q^j_{\vartheta'}(K^{\vartheta'})', \quad j = 2, \ldots, l;$$

moreover, we have the natural initial conditions $K^j(0) = g^j_2/2$. Hence the $K^j(\xi^1)$, $j = 2, \ldots, l$, are uniquely determined. The system

$$(L^j)' = P^j - \frac{K^j}{\xi^1} + Q^j_{\vartheta'}\frac{K^{\vartheta'}}{\xi^1} + N^j_{\vartheta'}L^{\vartheta'} + Q^j_{\vartheta'}(L^{\vartheta'})' \tag{4.26}$$

then determines the functions $L^j(\xi^1)$, $j = 2, \ldots, l$. It is true that we do not have initial conditions for these functions; however, the right sides in (4.26) have no singularities at the origin, and all the $L^j(\xi^1)$ are bounded on the interval $-\rho \leq \xi^1 \leq 0$.

We thus conclude that

$$\eta_1^j(\xi^1) = \frac{g_1^j}{2} \ln|\xi^1| + O(1), \quad j = 2, \ldots, l.$$

Here and in the sequel we write $O(1)$ to denote terms that are uniformly bounded for $-\rho \leq \xi^1 \leq 0$. Furthermore, (4.21) yields

$$\eta_1^1(\xi^1) = -\frac{1}{2\xi^1} - \frac{1}{2} b_{\varkappa'}^1 g_1^{\varkappa'} \ln|\xi^1| + O(1);$$

$$\xi_1^i(\xi^1) = \frac{M^i}{\xi^1} + N^i \ln|\xi^1| + O(1), \quad i = 2, \ldots, k,$$

where M^i and N^i are constants.

The functions $\xi_2^i(\xi^1)$, $i = 2,\ldots,k$, and $\eta_2^j(\xi^1)$, $j = 1,\ldots,l$, are calculated similarly. We do not reproduce the intermediate calculations, but give only the final result in a form that can be used for the calculation of the solution (4.11) up to and including terms of order ε:

$$\eta_2^1(\xi^1) = -\frac{1}{8(\xi^1)^4} + \delta^1(\xi^1),$$

$$\eta_2^j(\xi^1) = \delta^j(\xi^1), \quad j = 2, \ldots, l,$$

$$\xi_2^i(\xi^1) = \frac{q^i}{(\xi^1)^4} + \gamma^i(\xi^1), \quad i = 2, \ldots, k.$$

Here the q^i are numerical coefficients, and the functions $\delta^j(\xi^1)$, $j = 1,\ldots,l$, $\gamma^i(\xi^1)$, $i = 2,\ldots,k$, although unbounded when $\xi^1 \to 0$, are such that $\varepsilon^2\delta^j(\xi^1)$ and $\varepsilon^2\gamma^i(\xi^1)$ are uniformly of order $O(\varepsilon)$ for $\varepsilon \to 0$ on the interval $-\rho \leq \xi^1 \leq -\sigma_1$, where $\sigma_1 = \varepsilon^{2/7}$.

We have thus calculated the zeroth, the first, and the second formal approximations for the solution (4.11). The fact that each of these formal approximations actually represents the solution (4.11) on $-\rho \leq \xi^1 \leq -\sigma_1$ with a successively higher order of accuracy naturally requires a special proof. This proof is analogous [40] to the proof for the two-dimensional case developed in Chapter II, Sec. 7. Here we consider only the following result.

Theorem 3. On $-\rho \leq \xi^1 \leq -\varepsilon^{2/7}$, the solution (4.11) has the representation

$$\xi^i = \xi_0^i(\xi^1) + \varepsilon\xi_1^i(\xi^1) + \varepsilon^2\xi_2^i(\xi^1) + \mathfrak{R}^i(\xi^1, \varepsilon) \equiv$$

$$\equiv k^i(\xi^1)^2 + l^i(\xi^1)^3 + \ldots + \varepsilon\left(\frac{M^i}{\xi^1} + N^i \ln|\xi^1|\right) +$$

$$+ \varepsilon^2 \frac{q^i}{(\xi^1)^4} + \mathfrak{R}^i(\xi^1, \varepsilon), \quad i = 2, \ldots, k,$$

$$\eta^1 = \eta_0^1(\xi^1) + \varepsilon\eta_1^1(\xi^1) + \varepsilon^2\eta_2^1(\xi^1) + \mathfrak{S}^1(\xi^1, \varepsilon) \equiv$$
$$\equiv -(\xi^1)^2 + \left(\frac{2}{3} b^1_{\varkappa'} g_1^{\varkappa'} + c_1^1 - d_1^1 - r^1_{\alpha'} k^{\alpha'}\right)(\xi')^3 + \ldots \tag{4.27}$$
$$\ldots + \varepsilon\left(-\frac{1}{2\xi^1} - \frac{1}{2} b^1_{\varkappa'} g_1^{\varkappa'} \ln|\xi^1|\right) + \varepsilon^2 \frac{-1}{8(\xi^1)^4} + \mathfrak{S}^1(\xi^1, \varepsilon),$$
$$\eta^j = \eta_0^j(\xi^1) + \varepsilon\eta_1^j(\xi^1) + \mathfrak{S}^j(\xi^1, \varepsilon) \equiv$$
$$\equiv -\frac{2}{3} g_1^j(\xi^1)^3 + \ldots + \varepsilon \frac{g_1^j}{2} \ln|\xi^1| + \mathfrak{S}^j(\xi^1, \varepsilon), \quad j = 2, \ldots, l;$$

here the dots represent terms containing $(\xi^1)^4$ and higher powers, and the functions $\mathfrak{R}^i(\xi^1, \varepsilon)$, $i = 2, \ldots, k$, and $\mathfrak{S}^j(\xi^1, \varepsilon)$, $j = 1, \ldots, l$ are of order ε or higher on $-\rho \le \xi^1 \le -\varepsilon^{2/7}$.

5. Asymptotic Approximations for the Trajectory in the Neighborhood of a Junction Point

We now calculate the solution (4.11) for $-\sigma_1 \le \xi^1 \le \sigma_2$, where $\sigma_1 = \varepsilon^{2/9}$ (we recall that $\sigma_2 = \varepsilon^{2/7}$). We shall see that the deviation from the corresponding solution of the degenerate system is greater in this case; the deviations are of orders $\varepsilon^{2/3}$ and $\varepsilon \ln(1/\varepsilon)$.

We first make the variable change

$$\begin{aligned} \xi^1 &= \mu u^1, & \xi^i &= \mu^2 u^i, & i &= 2, \ldots, k, \\ \eta^1 &= \mu v^1, & \eta^j &= \mu^3 v^j, & j &= 2, \ldots, l, \\ t &= \mu^2 \tau, & \mu^3 &= \varepsilon \end{aligned} \tag{5.1}$$

(cf. Chapter II, Sec. 8); we consider μ as a new small parameter.

Here τ is a fast time, i.e., when $\mu \to 0$, a finite interval of the time t corresponds to an infinitely long interval of the time τ. Relations (5.1) establish a one-to-one correspondence between the small but finite neighborhood U_0 of the origin in the space of the variables $(\xi^1, \ldots, \xi^k, \eta^1, \ldots, \eta^l)$ and an infinitely large (for $\varepsilon \to 0$) region U_0^* of the space of the variables $(u^1, \ldots, u^k, v^1, \ldots, v^l)$; geometrically, they dilate along the ξ and η axes with different dilation coefficients, each of which increases with decreasing ε.

The substitution (5.1) transforms the system (4.1) into

$$\left\{\begin{aligned} &\dot{u}^1 = (u^1)^2 + v^1 + \mu\left[b^1_{\varkappa'} v^{\varkappa'} + c_1^1 u^1 v^1 + d_1^1 (u^1)^3 + r^1_{\alpha'} u^1 u^{\alpha'}\right] + \ldots, \\ &\mu\dot{u}^i = a^i_{\alpha'} u^{\alpha'} + b_1^i v^1 + c_0^i (u^1)^2 + \mu\left[b^i_{\varkappa'} v^{\varkappa'} + d_1^i (u^1)^3 + c_1^i u^1 v^1 + r^i_{\alpha'} u^1 u^{\alpha'}\right] + \ldots, \\ &\qquad\qquad i = 2, \ldots, k, \\ &\dot{v}^1 = 1 + \mu g_1^1 u^1 + \ldots, \\ &\dot{v}^j = g_1^j u^1 + \ldots, \qquad j = 2, \ldots, l \end{aligned}\right. \tag{5.2}$$

(differentiation with respect to τ is indicated by a dot), or, with new notation for the right sides in (5.2),

$$\begin{cases} \dot{u}^1 = \varphi^1(u^\alpha, v^\varkappa, \mu), \\ \mu\dot{u}^i = \varphi^i(u^\alpha, v^\varkappa, \mu), \quad i=2, \ldots, k, \\ \dot{v}^j = \psi^j(u^\alpha, v^\varkappa, \mu), \quad j=1, \ldots, l. \end{cases} \tag{5.3}$$

Taking u^1 for the independent variable, we replace (5.3) by the system

$$\begin{cases} \mu \dfrac{du^i}{du^1} = \dfrac{\varphi^i(u^\alpha, v^\varkappa, \mu)}{\varphi^1(u^\alpha, v^\varkappa, \mu)}, \quad i=2, \ldots, k, \\ \dfrac{dv^j}{du^1} = \dfrac{\psi^j(u^\alpha, v^\varkappa, \mu)}{\varphi^1(u^\alpha, v^\varkappa, \mu)}, \quad j=1, \ldots, l, \end{cases} \tag{5.4}$$

which is obtained from (4.10) by the substitution (5.1). It is natural to try to approximate the solution of (5.4) corresponding [under the transformation (5.1)] to the part of the solution (4.11) for $\xi^1 \geq -\sigma_1$ by the sums

$$\begin{aligned} u^{i,N} &= u_0^i(u^1) + \mu u_1^i(u^1) + \ldots + \mu^N u_N^i(u^1), \; i=2, \ldots, k, \\ v^{j,N} &= v_0^j(u^1) + \mu v_1^j(u^1) + \ldots + \mu^N v_N^j(u^1), \; j=1, \ldots, l, \end{aligned} \tag{5.5}$$

where the functions

$$u_\nu^i(u^1), v_\nu^j(u^1), \quad i=2, \ldots, k, \; j=1, \ldots, l, \; \nu=0, 1, \ldots, N,$$

are obtained from (5.4) by substitution of the expressions (5.5) and comparison of coefficients in expansions in powers of μ.

In this way we obtain differential equations for the functions under consideration; our first problem now is to choose completely definite solutions (the situation is analogous to that encountered in Chapter II, Sec. 10). We seek initial conditions for our special solutions by using compatibility conditions for approximations (5.5) to the functions (4.27) at the point $u^1 = -\sigma^1/\mu$ (i.e., $\xi^1 = -\sigma_1$) with the required degree of accuracy. We find, however, that almost all the functions $u_\nu^i(u^1)$ and $v_\nu^j(u^1)$ increase unboundedly when $u^1 \to +\infty$. Hence we do not try to approximate the solution to be calculated by sums of the form (5.5) on the whole interval $-\sigma_1 \leq \xi^1 \leq \rho$ (i.e., $-\sigma_1/\mu \leq u^1 \leq \rho/\mu$). We proceed as follows (Chapter II, Secs. 10 and 11): We put $\sigma_2 = \varepsilon^{\lambda_2}$, and select $\lambda_2 > 0$ so that we obtain a uniform asymptotic approximation with the required accuracy on the interval

$-\sigma_1/\mu \leq u^1 \leq \sigma_2/\mu$ (i.e., $-\sigma_1 \leq \xi^1 \leq \sigma_2$). We want to calculate $\mathfrak{T}_\varepsilon$ with accuracy up to order ε; further calculations show that it is sufficient to put $\sigma_2 = \varepsilon^{2/9}$ and find only the zeroth and first approximations.

We first calculate the functions $u_0^i(u^1)$, $i = 2,\ldots,k$, $v_0^j(u^1)$, $j = 1,\ldots,l$, describing the zeroth approximation. To this end we use the system of equations degenerate with respect to the system (5.4):

$$\begin{cases} a^i_{\alpha'} u^{\alpha'} + b^i_1 v^1 + c^i_0 (u^1)^2 = 0, & i = 2, \ldots, k, \\ \dfrac{dv^1}{du^1} = \dfrac{1}{(u^1)^2 + v^1}, & \\ \dfrac{dv^j}{du^1} = \dfrac{g^j_1 u^1}{(u^1)^2 + v^1}, & j = 2, \ldots, l. \end{cases} \tag{5.6}$$

In this system, the kth equation can be reduced independently to a Riccati equation (Chapter II, Sec. 9). For $v_0^1(u^1)$ we take the function (II.9.8) related to the separating solution of Riccati's equation. We recall the following formulas:

$$\begin{aligned} v_0^1(u^1) &= -(u^1)^2 + z_0(u^1), \quad -\infty < u^1 \leqslant 0, \\ z_0(u^1)^- &= -\frac{1}{2u^1} - \frac{1}{8(u^1)^4} + O\left(\frac{1}{(u^1)^7}\right), \\ v_0^1(u^1)^+ &= \Omega_0 - \frac{1}{u^1} + O\left(\frac{1}{(u^1)^3}\right), \quad \Omega_0 = \lim_{u^1 \to \infty} v_0^1(u^1). \end{aligned} \tag{5.7}$$

The last $l - 1$ equations (5.6) yield

$$v_0^j(u^1) = g_1^j \int \frac{u^1\,du^1}{(u^1)^2 + v_0^1(u^1)}, \quad j = 2, \ldots, l,$$

and we see that the $v_0^j(u^1)$ are obtained to within additive constants. We determine these functions as follows:

$$v_0^j(u^1) = g_1^j \int_0^{u^1} \frac{u^1\,du^1}{(u^1)^2 + v_0^1(u^1)} - \frac{g_1^j}{2} \ln \frac{1}{\mu}, \quad j = 2, \ldots, l. \tag{5.8}$$

Using (5.7) and performing some elementary transformations, we obtain the following asymptotic representations for the functions (5.8):

$$\begin{aligned} v_0^j(u^1)^- &= -\frac{2}{3} g_1^j (u^1)^3 + \frac{g_1^j}{2} \ln|u^1| - \frac{g_1^j}{2} \ln \frac{1}{\mu} + O(1), \\ v_0^j(u^1)^+ &= g_1^j \ln u^1 - \frac{g_1^j}{2} \ln \frac{1}{\mu} + O(1), \quad j = 2, \ldots, l. \end{aligned} \tag{5.9}$$

The first $k-1$ equations (5.6), which are finite equations, uniquely determine the functions $u_0^i(u^1)$, $i = 2,\ldots,k$, and after some calculations we find that

$$u_0^i(u^1)^- = k^i(u^1)^2 + \frac{M^i}{u^1} + \frac{q^i}{(u^1)^4} + O\left(\frac{1}{(u^1)^7}\right),$$
$$u_0^i(u^1)^+ = \Omega_0^i(u^1)^2 + s_0^i + \frac{n_0^i}{u^1} + O\left(\frac{1}{(u^1)^3}\right), \tag{5.10}$$
$$i = 2, \ldots, k,$$

where Ω_0^i, s_0^i, and n_0^i are numerical coefficients.

We now consider the calculation of the formal approximation

$$\begin{aligned} u^{i,1} &= u_0^i(u^1) + \mu u_1^i(u^1), \quad i = 2, \ldots, k, \\ v^{j,1} &= v_0^j(u^1) + \mu v_1^j(u^1), \quad j = 1, \ldots, l, \end{aligned} \tag{5.11}$$

i.e., the determination of the functions $u_1^i(u^1)$, $i = 2,\ldots,k$ and $v_1^j(u^1)$, $j = 1,\ldots,l$.

Using (5.11) in the right and left sides of the equation

$$\frac{dv^1}{du^1} = \frac{\psi^1(u^\alpha, v^\varkappa, \mu)}{\varphi^1(u^\alpha, v^\varkappa, \mu)}$$

and equating coefficients of μ in the expansions in powers of μ, we obtain the following linear differential equation for $v_1^1(u^1)$:

$$\frac{dv_1^1}{du^1} + \frac{v_1^1}{[(u^1)^2 + v_0^1(u^1)]^2} = \mathscr{H}_1(u^1) + \frac{b_{\varkappa'}^1 g_1^{\varkappa'}}{2\,[(u^1)^2 + v_0^1(u^1)]^2} \ln\frac{1}{\mu},$$

where

$$\mathscr{H}_1(u^1) = \frac{g_1^1 u^1 [(u^1)^2 + v_0^1(u^1)] - b_{\varkappa'}^1 \tilde{v}_0^{\varkappa'}(u^1)}{[(u^1)^2 + v_0^1(u^1)]^2} - \frac{c_1^1 u^1 v_0^1(u^1) + d_1^1(u^1)^3 + r_{\alpha'}^1 u^1 u_0^{\alpha'}(u^1)}{[(u^1)^2 + v_0^1(u^1)]^2},$$

$$\tilde{v}_0^{\varkappa'}(u^1) = g_1^{\varkappa'} \int_0^{u^1} \frac{u^1\,du^1}{(u^1)^2 + v_0^1(u^1)}, \quad \varkappa' = 2, \ldots, l.$$

This equation is independent of the functions $v_1^2,\ldots,v_1^l$, $u_1^2,\ldots,u_1^k$, and can be solved independently; we use the following particular solution:

$$v_1^1(u^1) = \mathscr{M}(u^1) \int_{-\infty}^{u^1} \frac{\mathscr{H}_1(\theta)}{\mathscr{M}(\theta)}\,d\theta + \frac{1}{2} b_{\varkappa'}^1 g_1^{\varkappa'} \ln\frac{1}{\mu}, \tag{5.12}$$

where

$$\mathcal{M}(u^1) = \exp \int_{u^1}^{\infty} \frac{d\theta}{[\theta^2 + v_0^1(\theta)]^2}$$

(II.10.11). By virtue of the representations (II.11.3), (II.11.4), (5.7), (5.9), and (5.10), formula (5.12) yields the following asymptotic representations for $v_1^1(u^1)$:

$$\begin{aligned} v_1^1(u^1)^- &= \left(\frac{2}{3} b^1_{\varkappa'} g_1^{\varkappa'} + c_1^1 - d_1^1 - r^1_{\alpha'} k^{\alpha'}\right)(u^1)^3 - \\ &- \frac{1}{2} b^1_{\varkappa'} g_1^{\varkappa'} \ln|u^1| + \frac{1}{2} b^1_{\varkappa'} g_1^{\varkappa'} \ln\frac{1}{\mu} + O(1), \\ v_1^1(u^1)^+ &= (g_1^1 - d_1^1 - r^1_{\alpha'} \Omega_0^{\alpha'}) \ln u^1 + \frac{1}{2} b^1_{\varkappa'} g_1^{\varkappa'} \ln\frac{1}{\mu} + O(1). \end{aligned} \tag{5.13}$$

When $v_1^1(u^1)$ is known we can uniquely determine the functions $u_1^i(u^1)$, $i = 2,\ldots,k$. To this end we use the sums (5.11) in both sides of the first $k - 1$ equations in (5.4) and equate coefficients of μ in the expansions in powers of μ:

$$\begin{aligned} (u_0^i)' [(u^1)^2 + v_0^1] &= a^i_{\alpha'} u_1^{\alpha'} + b_1^i v_1^1 + b^i_{\varkappa'} v_0^{\varkappa'} + \\ &+ d_1^i (u^1)^3 + c_1^i u^1 v_0^1 + r^i_{\alpha'} u^1 u_0^{\alpha'}, \quad i = 2, \ldots, k. \end{aligned}$$

This is a system of linear algebraic equations with a nonvanishing coefficient determinant, and it can be solved for the $u_1^i(u^1)$, $i = 2,\ldots,k$. We do not reproduce the resulting formulas; we only need asymptotic representations for large negative and large positive u^1. These representations are easily obtained by using (5.7), (5.9), (5.10), and (5.13):

$$\begin{aligned} u_1^i(u^1)^- &= l^i (u^1)^3 + N^i \ln|u^1| - N^i \ln\frac{1}{\mu} + O(1), \\ u_1^i(u^1)^+ &= \omega_1^i (u^1)^3 + s_1^i \ln u^1 + n_1^i \ln\frac{1}{\mu} + O(1), \\ & i = 2, \ldots, k, \end{aligned} \tag{5.14}$$

here ω_1^i, s_1^i, and n_1^i are numerical constants.

We finally have the following formal approximation for solutions of the system (5.4):

$$\begin{aligned} u^i &= u_0^i(u^1) + \mu u_1^i(u^1), \quad i = 2, \ldots, k, \\ v^1 &= v_0^1(u^1) + \mu v_1^1(u^1), \quad v^j = v_0^j(u^1), \quad j = 2, \ldots, l. \end{aligned} \tag{5.15}$$

We note that the functions $v_1^2(u^1),\ldots,v_1^l(u^1)$ (see 5.5) have not been found. This is because the approximation (5.15) ensures the required compatibility with the solution (4.27) at the point $\xi^1 = -\sigma_1$.

In fact, in the coordinates $\xi^1,\ldots,\xi^k,\eta^1,\ldots,\eta^l$, the approximation (5.15) can be written as follows [see (5.1)]:

$$\begin{aligned} \xi^i &= \varepsilon^{2/3}u_0^i(\xi^1/\mu)+\varepsilon u_1^i(\xi^1/\mu), \quad i=2,\ldots,k,\\ \eta^1 &= \varepsilon^{2/3}v_0^1(\xi^1/\mu)+\varepsilon v_1^1(\xi^1/\mu),\\ \eta^j &= \varepsilon v_0^j(\xi^1/\mu), \quad j=2,\ldots,l. \end{aligned} \tag{5.16}$$

Employing the asymptotic approximations of the functions

$$\begin{aligned} &u_0^i(u^1),\ u_1^i(u^1), && i=2,\ldots,k,\\ &v_0^1(u^1),\ v_1^1(u^1),\ v_0^j(u^1), && j=2,\ldots,l, \end{aligned} \tag{5.17}$$

for large negative u^1 [see (5.7), (5.9), (5.10), (5.13), and (5.14)], we can now find the values of the sums (5.16) for $\xi^1 = -\sigma_1$ with accuracy of order ε:

$$\xi^i = k^i\sigma_1^2 - l^i\sigma_1^3 + \varepsilon\left(-\frac{M^i}{\sigma_1}+N^i\ln\sigma_1\right)+\varepsilon^2\frac{q^i}{\sigma_1^4}+O(\varepsilon), \quad i=2,\ldots,k,$$

$$\begin{aligned} \eta^1 = -\sigma_1^2 - \left(\frac{2}{3}b^1_{\varkappa'}g_1^{\varkappa'}+c_1^1-d_1^1-r^1_{\alpha'}k^{\alpha'}\right)\sigma_1^3 + \\ +\varepsilon\left(\frac{1}{2\sigma_1}-\frac{1}{2}b^1_{\varkappa'}g_1^{\varkappa'}\ln\sigma_1\right)+\varepsilon^2\frac{-1}{8\sigma_1^4}+O(\varepsilon), \end{aligned}$$

$$\eta^j = \frac{2}{3}g_1^j\sigma_1^3+\varepsilon\frac{g_1^j}{2}\ln\sigma_1+O(\varepsilon), \quad j=2,\ldots,l.$$

Comparing these expressions with the expressions obtained from (4.27) for $\xi^1 = -\sigma_1$, where $\sigma_1 = \varepsilon^{2/7}$, we see that they actually coincide with accuracy up to quantities of order ε.

The fact that the formally constructed approximation (5.15) is actually an asymptotic approximation for solutions of (5.4) on $-\sigma_1/\mu \leq u^1 \leq \sigma_2/\mu$, where $\sigma_2 = \varepsilon^{2/9}$, with accuracy up to order ε, must be proved separately. The proof is analogous to the two-dimensional proof [40] developed in Chapter II, Sec. 11; we do not reproduce it.

Theorem 4. On the interval defined by the inequality $-\varepsilon^{2/7}\leqslant\xi^1\leqslant\varepsilon^{2/9}$ the solution (4.11) has the following representation:

$$\begin{aligned} \xi^i &= \varepsilon^{2/3}u_0^i(\xi^1/\mu)+\varepsilon u_1^i(\xi^1/\mu)+\mathfrak{R}^i(\xi^1,\varepsilon), \quad i=2,\ldots,k,\\ \eta^1 &= \varepsilon^{2/3}v_0^1(\xi^1/\mu)+\varepsilon v_1^1(\xi^1/\mu)+\mathfrak{S}^1(\xi^1,\varepsilon),\\ \eta^j &= \varepsilon v_0^j(\xi^1/\mu)+\mathfrak{S}^j(\xi^1,\varepsilon), \quad j=2,\ldots,l; \end{aligned} \tag{5.18}$$

here $\mu^3 = \varepsilon$, and the functions $\mathfrak{N}^i(\xi^1,\varepsilon)$, $i = 2,\ldots,k$ and $\mathfrak{S}^j(\xi^1,\varepsilon)$, $j = 1, \ldots, l$, are of order ε or lower on the whole interval $-\varepsilon^{2/7} \leqslant \xi^1 \leqslant \varepsilon^{2/9}$.

6. Asymptotic Approximation of a Trajectory at the End of a Junction Section

Here we find the solution (4.11) for $\sigma_2 \leq \xi^1 \leq \rho$, where ρ is a small nonzero number (we recall that $\sigma_2 = \varepsilon^{2/9}$). We shall see that, at the end of the junction part, i.e., at a finite distance from the surface Γ, the trajectory $\mathfrak{T}_\varepsilon$ can actually be represented as in (2.2).

For $\sigma_2 \leq \xi^1 \leq \rho$, the functions (4.11) form a solution of the nondegenerate system (4.10). Let

$$\xi^i = \tilde{\xi}_0^i(\xi^1), \quad i=2, \ldots, k, \quad \eta^j = \tilde{\eta}^j(\xi^1), \quad j=1, \ldots, l, \tag{6.1}$$

be the solution of the corresponding degenerate system

$$\begin{cases} \dfrac{d\xi^i}{d\xi^1} = \dfrac{\Phi^i(\xi^\alpha, \eta^\varkappa)}{\Phi^1(\xi^\alpha, \eta^\varkappa)}, & i=2, \ldots, k, \\ \dfrac{d\eta^j}{d\xi^1} = 0, & j=1, \ldots, l, \end{cases} \tag{6.2}$$

defined for $0 < \xi^1 \leq \rho$ and satisfying the initial conditions

$$\lim_{\xi^1 \to 0} \tilde{\xi}_0^i(\xi^1) = 0, \quad i=2, \ldots, k, \quad \tilde{\eta}_0^j(0) = 0, \quad j=1, \ldots, l;$$

the functions (6.1) describe the part of $\mathfrak{T}_0$ where it leaves the junction point (the origin). The system (6.2) directly implies that

$$\begin{aligned} \tilde{\xi}_0^i(\xi^1) &= \Omega_0^i(\xi^1)^2 + \ldots, & i &= 2, \ldots, k, \\ \tilde{\eta}_0^j(\xi^1) &= 0, & j &= 1, \ldots, l; \end{aligned} \tag{6.3}$$

the three dots denote terms of higher order in ξ^1, and the constants Ω_0^i [see (5.10)] satisfy the algebraic system

$$a_{\alpha'}^i \Omega_0^{\alpha'} + c_0^i = 0, \quad i=2, \ldots, k. \tag{6.4}$$

We shall attempt to approximate the solution (4.11) for $\xi^1 \geq \sigma_2$ as follows:

$$\begin{aligned} \tilde{\xi}^{i,3} &= \tilde{\xi}_0^i(\xi^1) + \tilde{\xi}_1^i(\xi^1)\varepsilon^{2/3} + \tilde{\xi}_2^i(\xi^1)\varepsilon \ln\frac{1}{\varepsilon} + \tilde{\xi}_3^i(\xi^1)\varepsilon, & i &= 2, \ldots, k, \\ \tilde{\eta}^{j,3} &= \tilde{\eta}_0^j(\xi^1) + \tilde{\eta}_1^j(\xi^1)\varepsilon^{2/3} + \tilde{\eta}_2^j(\xi^1)\varepsilon \ln\frac{1}{\varepsilon} + \tilde{\eta}_3^j(\xi^1)\varepsilon, & j &= 1, \ldots, l; \end{aligned} \tag{6.5}$$

here the functions

$$\tilde{\xi}^i_\nu(\xi^1),\quad \tilde{\eta}^j_\nu(\xi^1),\quad i=2,\ldots,k,\quad j=1,\ldots,l,\quad \nu=0,1,2,3,$$

are determined from the system (4.10) by substitution of the expressions (6.5) and comparison of coefficients in the expansions. We seek the initial conditions as compatibility conditions between the approximations (6.5) and the functions (5.18) at the point $\xi^1 = \sigma_2$ (i.e., $u^1 = \sigma_2/\mu$) with the required degree of accuracy. We want to find $\mathfrak{T}_\varepsilon$ with accuracy up to order ε, and further calculations show that it is sufficient to determine only the functions written in (6.5).

The functions $\tilde{\xi}^i_0(\xi^1)$ and $\tilde{\eta}^j_0(\xi^1)$ have already been found by solving the system (6.2), and have the representations (6.3). We shall calculate the functions $\tilde{\eta}^j_\nu(\xi^1)$, $j = 1,\ldots,l$, $\nu = 1, 2, 3$. First, using (5.18) and the asymptotic representations of the functions (5.17) for large positive u^1, we obtain the value of the solution (4.11) for $\xi^1 = \sigma_2$ with accuracy of order ε:

$$\begin{aligned}\xi^i &= \Omega^i_0\sigma_2^2 + s^i_0\varepsilon^{2/3} + \omega^i_1\sigma_2^3 + n^i_0\frac{\varepsilon}{\sigma_2} + \\ &+ \frac{s^i_1+n^i_1}{3}\varepsilon\ln\frac{1}{\varepsilon} + s^i_1\varepsilon\ln\sigma_2 + O(\varepsilon),\quad i=2,\ldots,k,\\ \eta^1 &= \Omega_0\varepsilon^{2/3} - \frac{\varepsilon}{\sigma_2} + \frac{1}{3}\left(g^1_1 - d^1_1 - r^1_{\alpha'}\Omega^{\alpha'}_0 + \frac{1}{2}b^1_{\varkappa'}g^{\varkappa'}_1\right)\varepsilon\ln\frac{1}{\varepsilon} + \\ &\qquad + (g^1_1 - d^1_1 - r^1_{\alpha'}\Omega^{\alpha'}_0)\varepsilon\ln\sigma_2 + O(\varepsilon),\\ \eta^j &= \frac{1}{6}g^j_1\,\varepsilon\ln\frac{1}{\varepsilon} + g^j_1\varepsilon\ln\sigma_2 + O(\varepsilon),\quad j=2,\ldots,l.\end{aligned} \tag{6.6}$$

Using the sums (6.5) in both sides of the last l equations in (4.10) and comparing coefficients of the first order in the expansions, we obtain the equations

$$\begin{aligned}&(\tilde{\eta}^1_1)' = 0,\quad (\tilde{\eta}^1_2)' = 0,\\ &(\tilde{\eta}^1_3)' = \frac{\Psi^1(\xi^1, \tilde{\xi}^{\alpha'}_0(\xi^1), \tilde{\eta}^{\varkappa}_0(\xi^1))}{\Phi^1(\xi^1, \tilde{\xi}^{\alpha'}_0(\xi^1), \tilde{\eta}^{\varkappa}_0(\xi^1))};\\ &(\tilde{\eta}^j_1)' = 0,\quad (\tilde{\eta}^j_2)' = 0,\\ &(\tilde{\eta}^j_3)' = \frac{\Psi^j(\xi^1, \tilde{\xi}^{\alpha'}_0(\xi^1), \tilde{\eta}^{\varkappa}_0(\xi^1))}{\Phi^1(\xi^1, \tilde{\xi}^{\alpha'}_0(\xi^1), \tilde{\eta}^{\varkappa}_0(\xi^1))},\quad j=2,\ldots,l.\end{aligned} \tag{6.7}$$

It follows from relations (6.3) and (3.13) that, for $\xi^1 \to +0$,

$$\frac{\Psi^1(\xi^1, \tilde{\xi}^{\alpha'}_0(\xi^1), \tilde{\eta}^{\varkappa}_0(\xi^1))}{\Phi^1(\xi^1, \tilde{\xi}^{\alpha'}_0(\xi^1), \tilde{\eta}^{\varkappa}_0(\xi^1))} = \frac{1}{(\xi^1)^2} + \frac{g^1_1 - d^1_1 - r^1_{\alpha'}\Omega^{\alpha'}_0}{\xi^1} + O(1),$$

$$\frac{\Psi^j(\xi^1, \tilde{\xi}^{\alpha'}_0(\xi^1), \tilde{\eta}^{\varkappa}_0(\xi^1))}{\Phi^1(\xi^1, \tilde{\xi}^{\alpha'}_0(\xi^1), \tilde{\eta}^{\varkappa}_0(\xi^1))} = \frac{g^j_1}{\xi^1} + O(1),\quad j=2,\ldots,l.$$

It is now clear that the solutions of Eqs. (6.7) on the interval $0 < \xi^1 \leq \rho$ can be written as

$$\begin{gathered}\tilde{\eta}_1^1(\xi^1)=K_1^1, \quad \tilde{\eta}_2^1(\xi^1)=K_2^1,\\ \tilde{\eta}_3^1(\xi^1)=-\frac{1}{\xi^1}+(g_1^1-d_1^1-r_{\alpha'}^1\Omega_0^{\alpha'})\ln\xi^1+O(1);\\ \tilde{\eta}_1^j(\xi^1)=K_1^j, \quad \tilde{\eta}_2^j(\xi^1)=K_2^j,\\ \tilde{\eta}_3^j(\xi^1)=g_1^j\ln\xi^1+O(1), \quad j=2,\ \ldots,\ l\end{gathered} \tag{6.8}$$

(Chapter II, Sec. 13). The constants K_1^j, K_2^j, $j = 1,\ldots,l$, must be chosen so that, for $j = 1,\ldots,l$, the value $\tilde{\eta}^{j,3}$ of the second sum in (6.5) for $\xi^1 = \sigma_2$ differs from the value of the component η^j of the solution (4.11) at $\xi^1 = \sigma_2$ [see (6.6)] by a quantity of order ε:

$$\begin{aligned}&K_1^1=\Omega_0, \quad K_2^1=\frac{1}{3}\left(g_1^1-d_1^1-r_{\alpha'}^1\Omega_0^{\alpha'}+\frac{1}{2}b_{\varkappa'}^1 g_1^{\varkappa'}\right);\\ &K_1^j=0, \quad K_2^j=\frac{1}{6}g_1^j, \quad j=2,\ \ldots,\ l.\end{aligned} \tag{6.9}$$

The functions $\tilde{\eta}_\nu^j(\xi^1)$, $j = 1,\ldots,l$, $\nu = 1, 2, 3$ are completely determined by (6.8) and (6.9).

We now consider the functions $\tilde{\xi}_\nu^i(\xi^1)$, $i = 2,\ldots,k$, $\nu = 1, 2, 3$. Using (6.5) in both sides of the first $k - 1$ equations of the system (4.10) and comparing coefficients of $\varepsilon^{2/3}$ in the expansions, we find that

$$(\Phi^1)^2(\tilde{\xi}_1^i)'=\left(\Phi^1\frac{\partial\Phi^i}{\partial\xi^{\beta'}}-\Phi^i\frac{\partial\Phi^1}{\partial\xi^{\beta'}}\right)\tilde{\xi}_1^{\beta'}+\left(\Phi^1\frac{\partial\Phi^i}{\partial\eta^\vartheta}-\Phi^i\frac{\partial\Phi^1}{\partial\eta^\vartheta}\right)\tilde{\eta}_1^\vartheta;$$

here the functions Φ^i, $i = 1,\ldots,k$, and their derivatives are calculated at the point $(\xi^1, \tilde{\xi}_0^{\alpha'}(\xi^1), \tilde{\eta}_0^\varkappa(\xi^1))$. It follows easily from (3.13), (6.3), (6.4), (6.7), and (6.8) that the functions $\tilde{\xi}_1^i(\xi^1)$, $i = 2,\ldots,k$ satisfy the following linear system with an irreducible singularity for $\xi^1 = 0$:

$$(\tilde{\xi}_1^i)'=\left[\frac{a_{\beta'}^i}{(\xi^1)^2}+\ldots\right]\tilde{\xi}_1^{\beta'}+\frac{b^i\Omega_0}{(\xi^1)^2}+\ldots, \quad i=2,\ \ldots,\ k;$$

the three dots stand for terms having poles of lower than second order for $\xi^1 = 0$. Linear systems of this type have been thoroughly investigated [10]. Similar linear systems are obtained for the functions $\tilde{\xi}_2^i(\xi^1)$, $\tilde{\xi}_3^i(\xi^1)$, $i = 2,\ldots,k$. Initial conditions for these systems must be chosen so that, for $i = 2,\ldots,k$, the value $\tilde{\xi}^{i,3}$ of the first sum in (6.5) for $\xi^1 = \sigma_2$ differs from the value of the component ξ^i of the solution (4.11) at $\xi^1 = \sigma_2$ [see (6.6)] by a quantity of order ε.

Of course it still remains to prove that the formally obtained approximation (6.5) is in fact an asymptotic approximation for the solution (4.11) on the interval $\sigma_2 \leq \xi^1 \leq \rho$ with accuracy up to quantities of order ε. This can be proved [40], but we do not reproduce the proof.

Theorem 5. For $\varepsilon^{2/9} \leq \xi^1 \leq \rho$, the solution (4.11) has the representation

$$\begin{aligned}
\xi^i &= \Omega_0^i (\xi^1)^2 + \ldots + \tilde{\xi}_1^i (\xi^1) \varepsilon^{2/3} + \tilde{\xi}_2^i (\xi^1) \varepsilon \ln \frac{1}{\varepsilon} + \\
&\quad + \tilde{\xi}_3^i (\xi^1) \varepsilon + \mathfrak{R}^i (\xi^1, \varepsilon), \quad i = 2, \ldots, k, \\
\eta^1 &= \Omega_0 \varepsilon^{2/3} + \frac{1}{3} \left(g_1^1 - d_1^1 - r_{\alpha'}^1 \Omega_0^{\alpha'} + \frac{1}{2} b_{\varkappa'}^1 g_1^{\varkappa'} \right) \varepsilon \ln \frac{1}{\varepsilon} + \\
&\quad + \varepsilon \left[-\frac{1}{\xi^1} + (g_1^1 - d_1^1 - r_{\alpha'}^1 \Omega_0^{\alpha'}) \ln \xi^1 \right] + \mathfrak{S}^1 (\xi^1, \varepsilon), \\
\eta^j &= \frac{1}{6} g_1^j \varepsilon \ln \frac{1}{\varepsilon} + \varepsilon g_1^j \ln \xi^1 + \mathfrak{S}^j (\xi^1, \varepsilon), \quad j = 2, \ldots, l,
\end{aligned} \tag{6.10}$$

where the functions $\mathfrak{R}^i (\xi^1, \varepsilon)$, $i = 2, \ldots, k$, and $\mathfrak{S}^j (\xi^1, \varepsilon)$, $j = 1, \ldots, l$ are of order ε or of higher order for $\varepsilon^{2/9} \leq \xi^1 \leq \rho$.

7. The Displacement Vector

Let $\Delta = (\Delta^1, \ldots, \Delta^l)$ be the vector with the components

$$\begin{aligned}
\Delta^1 &= \Omega_0 \varepsilon^{2/3} + \frac{1}{3} \left(g_1^1 - d_1^1 - r_{\alpha'}^1 \Omega_0^{\alpha'} + \frac{1}{2} b_{\varkappa'}^1 g_1^{\varkappa'} \right) \varepsilon \ln \frac{1}{\varepsilon}, \\
\Delta^j &= \frac{1}{6} g_1^j \varepsilon \ln \frac{1}{\varepsilon}, \quad j = 2, \ldots, l.
\end{aligned} \tag{7.1}$$

This vector will be called the *displacement vector* corresponding to the junction point $S(x_0, y_0)$. Comparison of (6.10) with (7.1) shows that Δ is, with accuracy of order ε, the deviation of the solution (4.11) at the point $\xi^1 = \rho$ from the linear space Σ^k of points $(\xi^1, \ldots, \xi^k, 0, \ldots, 0)$.

The system (4.1) is obtained from (1.1) by a linear transformation of coordinates in the neighborhood of the junction point $S(x_0, y_0)$. Under this transformation (see Sec. 3), the fast and slow variables remain distinct, i.e., the plane $X^k_{y_0^1, \ldots, y_0^l}$ is transformed into the subspace Σ^k and the plane $Y^l_{x_0^1, \ldots, x_0^k}$ is transformed into the subspace H^l consisting of the points $(0, \ldots, 0, \eta^1, \ldots, \eta^l)$. Hence Δ characterizes the deviation of the solution (4.11) at the end of the junction part from the plane $X^k_{y_0^1, \ldots, y_0^l}$.

The displacement vector depends on the junction point $S(x_0,y_0)$ but, naturally, does not depend on the choice of the local coordinate system in the neighborhood of this point. We shall obtain an invariant expression for Δ, i.e., we shall express its components in terms of quantities obtained directly from the right sides of (1.1) and the coordinates of the junction point.

First, it is clear that

$$\Delta = \overline{\Delta}^{\varkappa} h_{\varkappa}, \tag{7.2}$$

where $h_1,\ldots,h_l$ is the basis (3.6) and

$$\overline{\Delta}^1 = p^{-1/3} q^{-1/3} \Delta^1, \quad \overline{\Delta}^{\varkappa'} = \Delta^{\varkappa'}, \quad \varkappa' = 2, \ldots, l \tag{7.3}$$

[see (3.10)]. Employing (3.12) and (7.3) and using the Kronecker delta δ^i_j, we obtain the following expressions for the $\overline{\Delta}^j$ directly from (7.1):

$$\overline{\Delta}^j = \frac{1}{6p} \overline{g}^j_1 \varepsilon \ln \frac{1}{\varepsilon} + \delta^j_1 \frac{\Omega_0}{p^{1/3} q^{1/3}} \varepsilon^{2/3} + \\ + \delta^j_1 \left(\frac{\overline{b}^1_{\varkappa} \overline{g}^{\varkappa}_1}{6pq} - \frac{\overline{d}^1_1}{3p^2} - \frac{\overline{r}^1_{\alpha'} \Omega^{\alpha'}_0}{3p^{2/3} q^{2/3}} \right) \varepsilon \ln \frac{1}{\varepsilon}, \quad j = 1, \ldots, l.$$

Now using these expressions in (7.2), we obtain an invariant expression for the displacement vector,

$$\Delta = \frac{\Omega_0}{p^{1/3} q^{1/3}} g \varepsilon^{2/3} + \left[\frac{\hbar}{6p} + g \left(\frac{r}{6pq} - \frac{s}{3p^2} - \frac{k}{3p^{2/3} q^{2/3}} \right) \right] \varepsilon \ln \frac{1}{\varepsilon}. \tag{7.4}$$

We have already obtained invariant expressions for the vectors g and $\hbar$ and the numbers p, q, and $s = \overline{d}^1_1$ in (7.4); it only remains to find the constants

$$r = \overline{b}^1_{\varkappa} \overline{g}^{\varkappa}_1, \qquad k = \overline{r}^1_{\alpha'} \Omega^{\alpha'}_0.$$

The relations

$$\overline{b}^1_j = n_\alpha B^\alpha_\varkappa h^\varkappa_j, \qquad \overline{g}^\varkappa_1 h^j_\varkappa = C^j_\beta m^\beta$$

[see (3.9)] imply that

$$r = n_\alpha B^\alpha_\varkappa C^\varkappa_\beta m^\beta \tag{7.5}$$

We now calculate k. We have already obtained an expression for $\overline{r}^1_\alpha$, [see (3.9)], and the quantities $\Omega^{\alpha'}_0$ satisfy the linear system (6.4). Using the coordinate transformation formulas (3.7), we find that

$$c_0^{\beta'} = l_\nu^{\beta'} A_{\gamma\lambda}^\nu m^\gamma m^\lambda,$$

where the matrix $\| l_j^i \|$ is the inverse of $\| e_j^i \|$. Employing the expressions for $c_0^{\beta'}$ in the system (6.4) and solving the system for $\Omega_0^{\alpha'}$, we obtain

$$\Omega_0^{\alpha'} = -\, b_{\beta'}^{\alpha'} l_\nu^{\beta'} A_{\gamma\lambda}^\nu m^\gamma m^\lambda,$$

where $\| b_{\beta'}^{\alpha'} \|$ is the inverse of $\| a_{\beta'}^{\alpha'} \|$. Hence

$$k = -2 n_i A_{\alpha\beta}^i m^\alpha e_{\alpha'}^\beta b_{i'}^{\alpha'} l_\nu^{i'} A_{\gamma\lambda}^\nu m^\gamma m^\lambda.$$

By putting

$$\tilde{d}_\nu^\beta = e_{\alpha'}^\beta b_{i'}^{\alpha'} l_\nu^{i'};$$

we easily see that the matrix $\| \tilde{d}_\nu^\beta \|$ is uniquely determined by the matrix $\| A_\alpha^i \|$. Hence, finally,

$$k = -\frac{2}{p^{4/3} q^{2/3}} n_i A_{\alpha\beta}^i m^\alpha \tilde{d}_\nu^\beta A_{\gamma\lambda}^\nu m^\gamma m^\lambda. \tag{7.6}$$

CHAPTER V

SYSTEMS OF ARBITRARY ORDER. ALMOST-DISCONTINOUS PERIODIC SOLUTIONS

A degenerate system can have discontinuous periodic solutions. Under certain conditions there is, close to each closed trajectory corresponding to such a solution, at least one closed trajectory of the nondegenerate system; the periodic solution corresponding to this trajectory has the nature of a relaxation oscillation.

In this chapter we describe conditions ensuring the existence of a periodic solution, differing slightly from a discontinuous solution, of a system of arbitrary order with a small parameter ε multiplying certain derivatives. Using asymptotic methods, we calculate trajectories and their periods with accuracy up to order ε.

1. Auxiliary Results

We continue our study of phase trajectories of the nondegenerate system (IV.1.1); we retain all assumptions made in Chapter IV, Sec. 1. We also assume that the *degenerate system (IV.1.2) has a periodic solution* (Chapter I, Secs. 5 and 6), and we denote its trajectory by Z_0.

It follows from considerations in Chapter IV, Sec. 2 that Z_0 is a closed continuous curve in the space R^n consisting of a finite number of alternating slow- and fast-motion parts. Without loss of generality, we assume for definiteness that this trajectory is formed of four parts: two slow-motion parts and two fast-motion parts (Fig. 49). Let S_1 and S_2 be the junction points of Z_0 and let P_1 and P_2 be the corresponding subsequent drop points. We write $u_1(P_2,S_1)$ and $u_2(P_1,S_2)$ for the slow-motion parts, and $v_1(S_1,P_1)$ and $v_2(S_2,P_2)$ for the fast-motion parts. We recall that $u_1(P_2,S_1)$ and $u_2(P_1,S_2)$, except for their endpoints S_1 and S_2, are in the stable part Γ_- of the surface Γ, while S_1 and S_2 are in the set Γ_0.

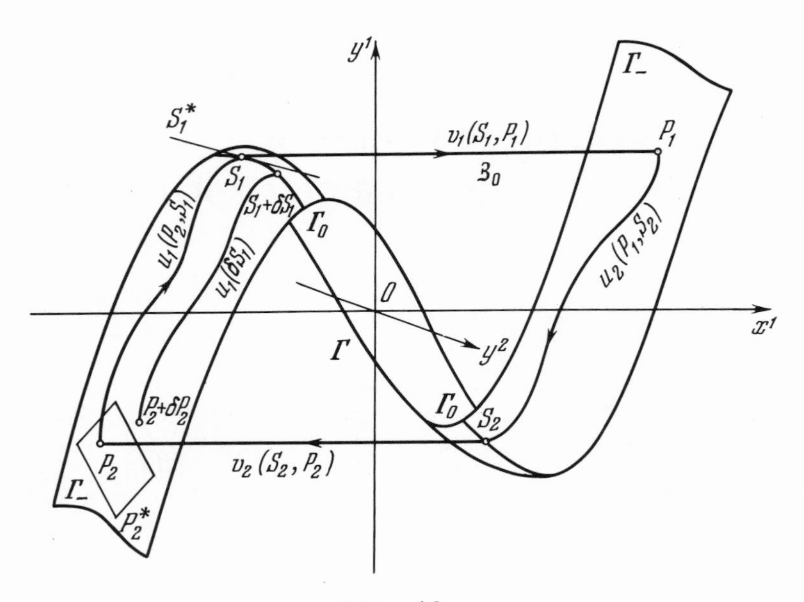

Fig. 49

We make the following assumptions concerning the discontinuous periodic solution of the degenerate system (IV.1.2): *The closed trajectory* Z_0 *is isolated and stable*. This means that a single circuit of the trajectory of the discontinuous solution of (IV.1.2) close to Z_0 generates a continuous mapping φ into itself of any sufficiently small but finite $(l-1)$-dimensional neighborhood of S_1 (or of S_2) consisting of points of Γ_0; the unique fixed point of this mapping φ is the point S_1 (or the point S_2). We also assume that the linear part of the mapping φ has the same property.

We shall need several auxiliary mappings induced by circuits of trajectories of discontinuous solutions of the degenerate system (IV.1.2).

Let S_1^* and S_2^* be the $(l-1)$-dimensional planes tangent, at the points S_1 and S_2, to the surface Γ_0 in R^n with the equations (IV.1.4) and (IV.1.6); let P_1^* and P_2^* be the l-dimensional planes tangent respectively at P_1 and P_2 to the surface Γ. We shall consider S_1^*, S_2^*, P_1^*, and P_2^* as vector spaces with origins at S_1, S_2, P_1, and P_2, respectively.

We consider the mapping

$$M_1^*\colon\ S_1^* \to P_2^*$$

of $\overset{*}{S}_1$ into $\overset{*}{P}_2$ defined as follows. Let t_1 be the time taken by the phase point of the degenerate system (IV.1.2) to traverse the part $u_1(P_2,S_1)$ of Z_0; let $S_1 + \delta S_1$ be any point on Γ_0 close to S_1. There is a point $P_2 + \delta P_2$ on Γ, close to the drop point P_2, which moves to the position $S_1 + \delta S_1$ during the same time t_1 on some trajectory $u_1(\delta S_1)$ of the system (IV.1.2) close to $u_1(P_2,S_1)$. The relation

$$S_1+\delta S_1 \rightarrow P_2+\delta P_2$$

thus determines a mapping

$$\mathfrak{M}\colon\ U(S_1)\rightarrow V(P_2)$$

of an $(l - 1)$-dimensional neighborhood $U(S_1)$ of S_1 consisting of points of the set Γ_0 into an l-dimensional neighborhood $V(P_2)$ of P_2 consisting of points of Γ. We denote the linear part of the mapping $\mathfrak{M}$ by M_1^*. The linear mapping

$$M_2^*\colon\ S_2^*\rightarrow P_1^*$$

is defined similarly.

We translate the spaces $\overset{*}{S}_1$, $\overset{*}{S}_2$, $\overset{*}{P}_1$, and $\overset{*}{P}_2$ so that their origins coincide with the origin of the space R^n, and then project them in the direction X^k onto the space Y^l. Under this projection π_x, the spaces P_1^* and P_2^* are mapped onto Y^l. In fact no degeneracy can be produced because the determinant of the matrix (IV.1.5) does not vanish at P_1 or P_2; hence the spaces P_1^* and P_2^* contain no directions parallel to the space X^k. Under the projection π_x, the spaces S_1^* and S_2^* are also mapped into Y^l without degeneracy, i.e., they are mapped onto $(l - 1)$-dimensional subspaces $\tilde{S}_1$ and $\tilde{S}_2$:

$$\pi_x S_1^*=\tilde{S}_1,\quad \pi_x S_2^*=\tilde{S}_2.$$

This is because, at the junction points S_1 and S_2, only one eigenvalue of the matrix (IV.1.5) vanishes.

The equations of the planes $\tilde{S}_1$ and $\tilde{S}_2$ can be obtained explicitly; as an example we shall find the equation of $\tilde{S}_1$. To this end it is sufficient to eliminate, from the equations

$$\frac{\partial f^i(S_1)}{\partial x^\alpha}(x^\alpha-x_1^\alpha)+\frac{\partial f^i(S_1)}{\partial y^\varkappa}(y^\varkappa-y_1^\varkappa)=0,\quad i=1,\ \ldots,\ k, \tag{1.1}$$

of the tangent plane to Γ at the point $S_1(x_1^1,\ldots,x_1^k,y_1^1,\ldots,y_1^l)$, the variables x^α, $\alpha = 1,\ldots,k$, and then replace $(y^\kappa - y_1^\kappa)$ by y^κ, $\kappa = 1,\ldots,l$. Let $n = (n_1,\ldots,n_k)$ be the eigenvector corresponding to the zero eigenvalue of the transpose of the matrix $\mathfrak{A}(S_1)$, and suppose that (IV.3.3) is satisfied. Forming the contraction of (1.1) with n_α, we obtain

$$n_\alpha B^\alpha_\varkappa (y^\varkappa - y^\varkappa_1) = 0; \quad B^\alpha_\varkappa = \frac{\partial f^\alpha (S_1)}{\partial y^\varkappa}.$$

Now consider the covariant vector

$$\begin{gathered} {}_1w = ({}_1w_1, \ldots, {}_1w_l), \\ {}_1w_j = n_\alpha B^\alpha_j, \quad j = 1, \ldots, l, \quad {}_1w_\varkappa g^\varkappa (S_1) = 1. \end{gathered} \tag{1.2}$$

The equation of $\tilde{S}_1$ can be written as

$${}_1w_\varkappa y^\varkappa = 0. \tag{1.3}$$

It can be proved similarly that $\tilde{S}_2$ has the equation

$${}_2w_\varkappa y^\varkappa = 0. \tag{1.4}$$

The projection π_x and the mappings M_1^* and M_2^* generate, in a natural fashion, the mappings

$$M_1 \colon \tilde{S}_1 \longrightarrow Y^l, \quad M_2 \colon \tilde{S}_2 \longrightarrow Y^l,$$

of $\tilde{S}_1$ and $\tilde{S}_2$ into Y^l

$$M_1 = \pi_x M_1^* \pi_x^{-1}, \quad M_2 = \pi_x M_2^* \pi_x^{-1}.$$

These mappings can be extended by various methods into mappings of Y^l onto itself. Let N_1 and N_2 be such extensions, under which the images of the vectors $g(S_1)$ and $g(S_2)$ are given by the formulas

$$N_1 g(S_1) = g(P_2), \quad N_2 g(S_2) = g(P_1).$$

We assume that the following *nondegeneracy condition* is satisfied: The vector $g(S_1)$ is transversal to the space S_1^*, and the vector $g(S_2)$ is transversal to the space S_2^*.

Hence we have mappings N_1 and N_2 of Y^l onto itself:

$$N_1 \colon Y^l \longrightarrow Y^l, \quad N_2 \colon Y^l \longrightarrow Y^l.$$

These mappings can be calculated by solving the system of equations in variations corresponding to the degenerate system (IV.1.2).

We use N_1 and N_2 to introduce the following linear mappings L_1 and L_2 of Y^l into itself:

$$\begin{aligned} L_1 y &\equiv N_1^{-1} y - ({}_1w, N_1^{-1} y)\, g(S_1), \\ L_2 y &\equiv N_2^{-1} y - ({}_2w, N_2^{-1} y)\, g(S_2). \end{aligned}$$

We shall prove that L_1 transforms Y^l into $\tilde{S}_1$, and L_2 transforms Y^l into $\tilde{S}_2$. Clearly, $Y^l = Y_1 + Y_2$, where $Y_1 = N_1\tilde{S}_1$ and Y_2 is the one-dimensional subspace spanned by the vector $g(P_2)$. If $y \in Y_1$, then

$$N_1^{-1}y \in \tilde{S}_1, \quad ({}_1w, N_1^{-1}y) = 0,$$

and $L_1 y \in \tilde{S}_1$. If $y \in Y_2$, then

$$y = \lambda g(P_2), \qquad N_1^{-1}y = \lambda g(S_1),$$

and $L_1 y = 0$, by virtue of the last of relations (1.2). The reasoning in the case of L_2 is similar.

We also define mappings Π_1 and Π_2 of the spaces $\tilde{S}_1$ and $\tilde{S}_2$ on themselves as the compositions of the mappings L_1 and L_2:

$$\Pi_1 = L_1 L_2, \qquad \Pi_2 = L_2 L_1. \tag{1.5}$$

It is easily seen that the mappings (1.5) are induced by the projections on Y^l of the linear approximations of the mappings of the neighborhoods $U(S_1)$ and $U(S_2)$ into themselves determined by the passage along phase trajectories of the degenerate system (IV.1.2). The unique fixed point of the mappings (1.5) is the vector $y = 0$.

Finally, we use the mappings (1.5) to construct mappings Π_1^ε and Π_2^ε of the spaces $\tilde{S}_1$ and $\tilde{S}_2$ onto themselves as follows:

$$\begin{aligned} \Pi_1^\varepsilon y &\equiv L_1 [L_2 (y + \Delta_1) + \Delta_2], \\ \Pi_2^\varepsilon y &\equiv L_2 [L_1 (y + \Delta_2) + \Delta_1]; \end{aligned} \tag{1.6}$$

here Δ_1 and Δ_2 are the displacement vectors corresponding to the points S_1 and S_2 (see Chapter IV, Sec. 7). The mappings (1.6) have been constructed formally. We shall use these mappings in our study of an almost-discontinuous periodic solution of (IV.1.1).

2. The Existence of an Almost-Discontinuous Periodic Solution. Asymptotic Calculation of the Trajectory

Consider the following question: Does the nondegenerate system (IV.1.1), for small ε, have a closed trajectory having the zeroth asymptotic approximation Z_0? It turns out that, under the assumptions made above concerning the system (IV.1.1), there is at least one such trajectory. However, in contrast to the two-dimensional case (Chapter III, Sec. 1), this is not easily proved. Moreover, it is not yet known whether such a trajectory is unique.

We shall prove that the nondegenerate system (IV.1.1) has a periodic solution whose trajectory Z_ε tends uniformly, for $\varepsilon \to 0$, to the closed curve Z_0, and simultaneously calculate the trajectory Z_ε with accuracy up to small quantities of order ε.

Theorem 1. Let $\delta \in \tilde{S}_1$ be the fixed point of the mapping Π_1^ε, and let $\delta_2 \in S_2$ be the fixed point of Π_2^ε:

$$\begin{aligned} \delta_1 &= L_1[L_2(\delta_1+\Delta_1)+\Delta_2], \\ \delta_2 &= L_2[L_1(\delta_2+\Delta_2)+\Delta_1]. \end{aligned} \tag{2.1}$$

Suppose that $\delta_1 S_1$ and $\delta_2 S_2$ are the inverse images of δ_1 and δ_2 under the projection onto the tangential planes S_1^* and S_2^*, and let $S_1 + \delta S_1$ and $S_2 + \delta S_2$ be the points of the neighborhoods $U(S_1)$ and $U(S_2)$ on Γ_0 corresponding to $\delta_1 S_1$ and $\delta_2 S_2$. Finally, let $u_1(\delta S_1)$ and $u_2(\delta S_2)$ be the parts of trajectories of the degenerate system (IV.1.2) passing respectively close to the parts $u_1(P_2,S_1)$ and $u_2(P_1,S_2)$ and abutting the junction points $S_1 + \delta S_1$ and $S_2 + \delta S_2$.

Then the nondegenerate system (IV.1.1) has a periodic solution Z_ε which, during a finite interval of time t, passes into ε-neighborhoods of the parts $u_1(\delta S_1)$ and $u_2(\delta S_2)$.

To prove this theorem we need the following lemma, which is also of independent interest.

Lemma 1. Let $X_{y_0}^k$ be the k-dimensional plane in R^n formed of the points (x,y_0), where y_0 is a constant vector. Let (x_0,y_0) be the point of intersection of $X_{y_0}^k$ with Γ. Assume that, at this point, all eigenvalues of the matrix $\mathfrak{A}$ have negative real parts, and consider the two solutions

$$x = x_1(t), \qquad y = y_1(t), \tag{2.2}$$

$$x = x_2(t), \qquad y = y_2(t), \tag{2.3}$$

of the system (IV.1.1), the first passing for $t = t_0$ through (x_0,y_0) and the second, for $t = t_0$, having its initial point in the plane $X_{y_0}^k$ at a nonzero but sufficiently small distance from (x_0,y_0).

Then there is a number $t' > t_0$ such that, for $t \geq t'$, the solutions (2.2) and (2.3) coincide to within quantities of order ε, for some finite time interval:

$$x_2(t) = x_1(t) + O(\varepsilon), \qquad y_2(t) = y_1(t) + O(\varepsilon).$$

It can be assumed that $|t' - t_0| \to 0$ when $\varepsilon \to 0$.

This lemma is essentially a particular case of Theorem 2, Chapter IV [34, 14]. We shall give a simple direct proof using the Lyapunov function method.

The point (x_0, y_0) is a point of exponentially stable equilibrium for the following system:

$$\frac{dx^i}{d\theta} = f^i(x^1, \ldots, x^k, y_0^1, \ldots, y_0^l), i = 1, \ldots, k, \ \theta = t/\varepsilon \tag{2.4}$$

[see (IV.1.3)]. It is known [45] that, in this case, there is a positive definite quadratic form $W(u)$ in the components of $u = x - x_0$, called a *Lyapunov function*. The derivative of this function calculated in accordance with the system obtained by linearizing (2.4) satisfies the inequality

$$\frac{\partial W(u)}{\partial u^i} \frac{\partial f^i(x_0, y_0)}{\partial x^j} u^j < -\alpha W(u), \tag{2.5}$$

where α is a positive constant. We put

$$u(t) = x_2(t) - x_1(t), \quad v(t) = y_2(t) - y_1(t),$$

and calculate the derivative of $W(u(t))$ with respect to t,

$$\frac{d}{dt}[W(u(t))] = \frac{\partial W}{\partial u^i} \frac{d(x_2^i - x_1^i)}{dt} = \frac{1}{\varepsilon} \frac{\partial W}{\partial u^i} [f^i(x_2, y_2) - f^i(x_1, y_1)] =$$

$$= \frac{1}{\varepsilon} \frac{\partial W}{\partial u^i} [f^i(x_2, y_2) - f^i(x_1, y_2) + f^i(x_1, y_2) - f^i(x_1, y_1)] =$$

$$= \frac{1}{\varepsilon} \frac{\partial W}{\partial u^i} \left[\frac{\partial f^i}{\partial x^j} (x_2^j - x_1^j) + \frac{\partial f^i}{\partial y^\nu} (y_2^\nu - y_1^\nu) \right];$$

the derivatives $\partial f^i / \partial x^j$ and $\partial f^i / \partial y^\nu$, as in the mean-value theorem, are evaluated for intermediate values of the arguments. By virtue of (2.5)

$$\frac{d}{dt}[W(u(t))] < -\frac{1}{\varepsilon} \tilde{\alpha} W(u(t)) + \frac{1}{\varepsilon} \frac{\partial W}{\partial u^i} \frac{\partial f^i}{\partial y^\nu} (y_2^\nu - y_1^\nu), \tag{2.6}$$

$$\tilde{\alpha} = \text{const} > 0.$$

The norms $\|u\|$, and $\|v\|$ of $u = x_2 - x_1$ and $v = y_2 - y_1$ are defined to be

$$\|u\| = \sqrt{W(u)}, \quad \|v\| = \sqrt{\sum_{\nu=1}^{l} (y_2^\nu - y_1^\nu)^2}.$$

Differentiation of the quantity $\|u(t)\|^2 = W(u(t))$ with respect to t, the use of (2.6), and application of elementary transformations yield

$$\frac{d}{dt}\|x_2(t)-x_1(t)\| < -\frac{1}{\varepsilon}a\|x_2(t)-x_1(t)\| + \frac{1}{\varepsilon}b\|y_2(t)-y_1(t)\|,$$

$$a=\text{const}>0, \quad b=\text{const}>0.$$

An estimate of the derivative of $\|v(t)\|$ with respect to t can be obtained similarly:

$$\frac{d\|v(t)\|}{dt} = \frac{1}{\|v\|}\sum_{j=1}^{l}(y_2^j-y_1^j)\frac{d}{dt}(y_2^j-y_1^j) < q\sum_{j=1}^{l}\left|\frac{d}{dt}(y_2^j-y_1^j)\right|, \quad q=\text{const}>0.$$

Since

$$\frac{d}{dt}(y_2^j-y_1^j) = g^j(x_2, y_2) - g^j(x_1, y_2) + g^j(x_1, y_2) - g^j(x_1, y_1) =$$
$$= \frac{\partial g^j}{\partial x^i}(x_2^i - x_1^i) + \frac{\partial g^j}{\partial y^\nu}(y_2^\nu - y_1^\nu), \quad j=1, \ldots, l,$$

we conclude that

$$\frac{d\|v\|}{dt} < c(\|u\|+\|v\|), \quad c=\text{const}>0.$$

Hence, putting

$$\xi=\|u\|, \quad \eta=\|v\|, \tag{2.7}$$

we obtain the differential inequalities

$$\begin{cases} \dfrac{d\xi}{dt} < -\dfrac{1}{\varepsilon}a\xi + \dfrac{1}{\varepsilon}b\eta, \\ \dfrac{d\eta}{dt} < c(\xi+\eta); \end{cases} \tag{2.8}$$

here a, b, and c are positive constants, and we can clearly assume that $b > a$. Together with the system (2.8) we consider the system of linear differential equations obtained by replacing the inequality signs in (2.8) by equality signs:

$$\begin{cases} \dfrac{d\xi}{dt} = -\dfrac{1}{\varepsilon}a\xi + \dfrac{1}{\varepsilon}b\eta, \\ \dfrac{d\eta}{dt} = c(\xi+\eta). \end{cases} \tag{2.9}$$

We shall compare the functions ξ and η given by (2.7) and satisfying the inequality system (2.8) with solutions of Eqs. (2.9). To this end we consider the triangle OAB (Fig. 50) in the (ξ,η) plane with the sides

$$OA:\ \eta=0, \quad OB:\ -a\xi+b\eta=0, \quad AB:\ \eta+2\varepsilon\frac{c}{a}(\xi-1)=0,$$

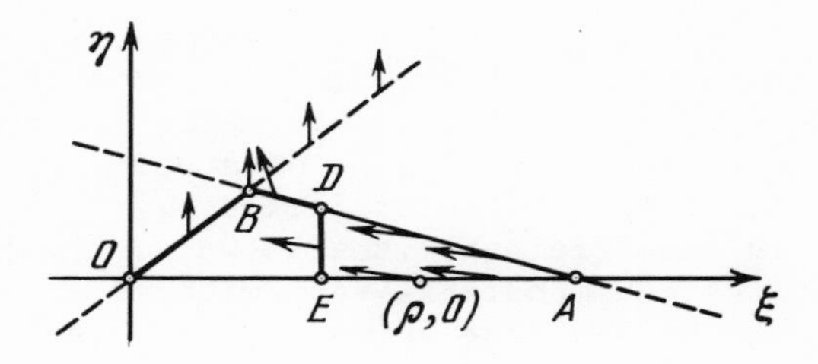

Fig. 50

and prove that, through each point of AB with an abscissa $\xi > \xi_0$, where ξ_0 is a small positive number of order ε, a trajectory of (2.9) enters the interior of OAB. In fact, by virtue of (2.9), the derivative of the function $\eta + 2\varepsilon(c/a)(\xi - 1)$, at any point of AB, is equal to

$$V(\xi) = \left\{c(\xi+\eta) + 2\varepsilon\frac{c}{a}\left(-\frac{1}{\varepsilon}a\xi + \frac{1}{\varepsilon}b\eta\right)\right\}_{\eta=-2\varepsilon\frac{c}{a}(\xi-1)}.$$

It can be verified directly that this derivative is negative for all $\xi > \xi_0$, where ξ_0 satisfies $V(\xi) = 0$ and, as is easily verified, is of order ε.

Besides the triangle OAB, we consider the triangle ADE with a vertex $E(\xi_0, 0)$ and the side DE parallel to the $O\eta$ axis (see Fig. 50). Let $(\rho, 0)$ be any point on AE. Elementary reasoning shows that the solution of the system (2.9) starting, for $t = t_0$, at this point cuts DE for some $t' > t_0$, and $t' \to t_0$ when $\varepsilon \to 0$. Hence it follows from (2.8) that, if the curve $\xi = \xi(t)$, $\eta = \eta(t)$ [where $\xi(t)$ and $\eta(t)$ are given by (2.7)] starts for $t = t_0$ at the point $(\rho, 0)$, then it cuts DE for some $t'' < t'$.

This means, firstly, that the norm $\|x_2(t) - x_1(t)\|$ after the time $t'' - t_0$ decreases to a value of order ε, and, secondly, that during the same time interval, the norm $\|y_2(t) - y_1(t)\|$ increases by a quantity of order ε at most. The relations

$$\|x_2(t) - x_1(t)\| = O(\varepsilon), \quad \|y_2(t) - y_1(t)\| = O(\varepsilon)$$

thus hold on some finite interval of variation of t. This proves the lemma.

We now prove Theorem 1. We first note that the vectors δ_1 and δ_2 [see (2.1)] exist by virtue of our assumption concerning the stability of the closed trajectory Z_0.

Take any point q on the part $u_1(\delta S_1)$ at a finite distance from the junction point $S_1 + \delta S_1$, and let $V^{l-1}(q)$ be a sufficiently small (but of finite size independent of ε) $(l - 1)$-dimensional neighborhood of a point on Γ, transversal at this point to the trajectory $u_1(\delta S_1)$. Suppose that $W^k(q)$ is a sufficiently small (but of finite side independent of ε) k-dimensional cube with edges parallel to the coordinate axes of the space X^k, with center at q. We write $\mathcal{O}^{k+l-1}(q)$ for the cartesian product $V^{l-1}(q) \times W^k(q)$.

Let u^ε be a trajectory of (IV.1.1) with initial point q^ε at a distance from q of order ε, i.e., $\rho(q,q^\varepsilon) = O(\varepsilon)$. We prove first that the trajectory u^ε again intersects $\mathcal{O}^{k+l-1}(q)$ at some point Q^ε, and that $\rho(q,Q^\varepsilon) = O(\varepsilon)$.

In the light of the results in Chapter IV, some finite part of the trajectory u^ε is along the part $u_1(\delta S_1)$, and the trajectory deviates from this part only by a quantity of order ε. However, near the point $S_1 + \delta S_1$, the deviation of u^ε from $u_1(\delta S_1)$ becomes of order $a\varepsilon^{2/3} + b\varepsilon \ln (1/\varepsilon)$.

We shall write $X^k_{S_1+\delta S_1}$ for the k-dimensional plane in R^n, parallel to the subspace X^k and passing through the junction point $S_1 + \delta S_1$. This plane contains the continuation of the trajectory $u_1(\delta S_1)$ of the system (IV.1.2). The results in Chapter IV imply that the trajectory u^ε does not pass through the point $S_1 + \delta S_1$, and continues, with accuracy up to quantities of order ε, in the k-dimensional plane $y =$ const in R^n obtained by translating the plane $X^k_{S_1+\delta S_1}$ by means of the displacement vector $\Delta(S_1 + \delta S_1)$ corresponding to the point $S_1 + \delta S_1$. But it follows from (2.1) that the vector δ_1 is of order $\varepsilon^{2/3}$; hence S_1 differs from $S_1 + \delta S_1$ by a quantity of order $\varepsilon^{2/3}$; thus

$$\Delta (S_1 + \delta S_1) = \Delta_1 + O (\varepsilon).$$

We can therefore assume that the extension of the trajectory u^ε, to within quantities of order ε, is in the plane $X^k_{S_1+\delta S_1}(\Delta_1)$ in R^n obtained by translating $X^k_{S_1+\delta S_1}$ by the vector Δ_1.

Let $P_1 + \delta P_1$ be a point of intersection of the plane $X^k_{S_1+\delta S_1}(\Delta_1)$ with Γ, and consider the trajectory of (IV.1.2) passing through this point. Assume that this trajectory has the junction point $S_2 + \delta S_2$; the trajectory can then be denoted by $u_2(\delta S_2)$. The image of the deviation S_2 from $S_2 + \delta S_2$ in the tangent plane S_2^* will be denoted by $\delta_1 S_2$, and the image of the vector $\delta_1 S_2$ under projection onto the space Y^l

will be denoted by δ_2. It is easily seen that

$$\delta_2 = L_2(\delta_1 + \Delta_1).$$

In its approach to the surface Γ (i.e., to the point $P_1 + \delta P_1$), the trajectory u^ε will leave the plane $X^k_{S_1+\delta S_1}(\Delta_1)$; however, Lemma 1 implies that some finite part of the trajectory will be in the ε-neighborhood of the trajectory $u_2(\delta S_2)$. This coincidence of u^ε with $u_2(\delta S_2)$ to within quantities of order ε is violated only near the point $S_2 + \delta S_2$; this point is avoided by the trajectory u^ε, which deviates from $u_2(\delta S_2)$ by quantities of order $a\varepsilon^{2/3} + b\varepsilon \ln(1/\varepsilon)$. The trajectory then returns to the ε-neighborhood of a part $u_1(\delta S_3)$ of some trajectory of the system (IV.1.2). The foregoing reasoning implies that the image of the vector $\delta_1 S_3$ in the space $\tilde{S}_3$ is

$$L_1[L_2(\delta_1 + \Delta_1) + \Delta_2],$$

i.e., its image is the vector δ_1 [see (2.1)]. Hence the part $u_1(\delta S_3)$ coincides, to within quantities of order ε, with the part $u_1(\delta S_1)$.

To complete the proof of the theorem, it remains to show that there is a trajectory u^ε with the behavior described above, identical to the closed trajectory Z_ε. To this end, it is sufficient to establish that circuits of trajectories of the system (IV.1.1) determine a continuous mapping of the set $\mathfrak{G}^{k+l-1}(q)$ *into itself*. But this follows directly from the above reasoning and Lemma 1, when it is taken into account that a circuit on a trajectory of a discontinuous solution of (IV.1.2) generates a continuous mapping of the neighborhood $V^{l-1}(q)$ into itself (by virtue of our assumption concerning the stability of the trajectory Z_0).

3. An Asymptotic Formula for the Period of Relaxation Oscillations

Theorem 1 asserts that, for sufficiently small ε, the system (IV.1.1) has a periodic solution which tends uniformly to a discontinuous periodic solution of the system (IV.1.2) when $\varepsilon \to 0$. The trajectory Z_ε of a periodic solution of (IV.1.1) was calculated with accuracy up to quantities of order ε. We shall calculate the period of the solution with the same accuracy.

Suppose that the part $u_1(P_2, S_1)$ is traversed by the phase point of the system (IV.1.2) in the time T_{10}, and the part $u_2(P_1,S_2)$ is traversed in the time T_{20}, so that the period of the discontinuous periodic solution with the trajectory Z_0 is

$$T_0 = T_{10} + T_{20}. \tag{3.1}$$

In addition to the parts $u_1(P_2,S_1)$ and $u_2(P_1,S_2)$, we consider the following parts of trajectories of the system

$$u_1(\delta S_1) = u_1(P_2 + \delta P_2,\ S_1 + \delta S_1),$$
$$u_2(\delta S_2) = u_2(P_1 + \delta P_1,\ S_2 + \delta S_2);$$

these parts are close to u_1 and u_2, respectively, and have the junction points $S_1 + \delta S_1$ and $S_2 + \delta S_2$, where the deviations δS_1 and δS_2 are determined by Theorem 1. The initial points $P_2 + \delta P_2$ and $P_1 + \delta P_1$ are chosen so that the time of motion of the phase point of the system (IV.1.2) on $u_1(\delta S_1)$ is T_{10}, and on $u_2(\delta S_2)$ it is T_{20}. We note that the deviations δS_1 and δS_2 are of order $\varepsilon^{2/3}$. We now shorten $u_1(\delta S_1)$ and $u_2(\delta S_2)$, that is, we replace them by their parts

$$\tilde{u}_1(\delta S_1) = \tilde{u}_1(P_2', S_1'), \quad \tilde{u}_2(\delta S_1) = \tilde{u}_2(P_1', S_2'),$$

whose initial and final points P_2', S_1', P_1', and S_2' are at sufficiently small but positive distances from the initial and final points of the parts $u_1(\delta S_1)$ and $u_2(\delta S_2)$.

Let the part $\tilde{u}_1(\delta S_1)$ be described by the following solution of the degenerate system

$$x^i = x_0^i(t), \quad i = 1, \ldots, k, \qquad y^j = y_0^j(t), \quad j = 1, \ldots, l;$$

let $t_1 \leq t \leq t_2$. It is clear that the time of motion of the phase point of the system (IV.1.2) on the part $\tilde{u}_1(\delta S_1)$ is $t_2 - t_1$. By virtue of Theorem 1, the trajectory Z_ε contains a part that can be represented as

$$x^i = x_0^i(t) + x_1^i(t, \varepsilon), \quad i = 1, \ldots, k,$$
$$y^j = y_0^j(t) + y_1^j(t, \varepsilon), \quad j = 1, \ldots, l,$$

for $t_1 \leq t \leq t_2$, where the functions $x_1^i(t,\varepsilon)$ and $y_1^j(t,\varepsilon)$ are of order ε for $t_1 \leq t \leq t_2$. We denote this part by $Z_\varepsilon(\tilde{u}_1(\delta S_1))$; the time of motion of the phase point of the system (IV.1.1) on the part $Z_\varepsilon(\tilde{u}_1(\delta S_1))$ is $t_2 - t_1 + O(\varepsilon)$.

Let $Z_\varepsilon(S_1 + \delta S_1)$ be the part of Z_ε directly following $Z_\varepsilon(\tilde{u}_1(\delta S_1))$ and passing around the junction point $S_1 + \delta S_1$; we assume that the part $Z_\varepsilon(S_1 + \delta S_1)$ has a small but nonzero length. In the neighborhood of $S_1 + \delta S_1$, the system (IV.1.1) can be reduced by a linear coordinate transformation to the canonical form (IV.3.11) and, on the part $Z_\varepsilon(S_1 + \delta S_1)$, we can use ξ^1 as the independent variable. Then the equation of this part becomes

$$\xi^i = \xi^i(\xi^1, \varepsilon),\ i = 2, \ldots, k,\ \eta^j = \eta^j(\xi^1, \varepsilon),\ j = 1, \ldots, l, \quad -\rho \leqslant \xi^1 \leqslant \rho. \tag{3.2}$$

The functions in (3.2) were calculated in Chapter IV, Secs. 4-6 with accuracy up to small quantities of order ε. Using these calculated values, we obtain, with the same accuracy, the time $T_{-\rho,\rho}$ for the passage of the phase point of (IV.1.1) along the part $Z_\varepsilon(S_1 + \delta S_1)$.

Now divide the interval $-\rho \leq \xi^1 \leq \rho$ by the points $\xi^1 = -\sigma_1$, $\xi^1 = 0$, and $\xi^1 = \sigma_2$, where $\sigma_1 = \varepsilon^{2/7}$ and $\sigma_2 = \varepsilon^{2/9}$, into four intervals $-\rho \leq \xi^1 \leq -\sigma_1$, $-\sigma_1 \leq \xi^1 \leq 0$, $0 \leq \xi^1 \leq \sigma_2$, and $\sigma_2 \leq \xi^1 \leq \rho$. Clearly

$$T_{-\rho,\rho} = T_{-\rho,-\sigma_1} + T_{-\sigma_1,0} + T_{0,\sigma_2} + T_{\sigma_2,\rho}\,. \tag{3.3}$$

In calculating the time, we start from the formula

$$T_{-\rho,\rho} = \int_{-\rho}^{\rho} \frac{d}{d\xi^1}[\eta^1(\xi^1, \varepsilon)]\,\delta(\xi^1, \xi^{\alpha'}(\xi^1, \varepsilon), \eta^\varkappa(\xi^1, \varepsilon))\,d\xi^1, \tag{3.4}$$

where $\xi^{\alpha'}(\xi^1,\varepsilon)$, $\eta^\kappa(\xi^1,\varepsilon)$, $\alpha' = 2,\ldots,k$, $\kappa = 1,\ldots,l$, are the functions (3.2), and

$$\delta(\xi^1, \xi^{\alpha'}, \eta^\varkappa) \equiv \frac{1}{\Psi^1(\xi^1, \xi^{\alpha'}, \eta^\varkappa)}$$

[see (IV.3.13)]. We have the asymptotic representations (IV.4.27) for the functions (3.2) on the interval $-\rho \leq \xi^1 \leq -\sigma_1$. We consider the integral in (3.4) on the interval $-\rho \leq \xi^1 \leq -\sigma_1$; we use the representations (IV.4.27) in the integrand, and we employ the Taylor expansion; we note, in particular, that

$$\delta(\xi^1, \xi_0^{\alpha'}(\xi^1), \eta_0^\varkappa(\xi^1)) = 1 - g_1^1\xi^1 + O((\xi^1)^2)$$

when $\xi^1 \to 0$.

We find, by applying simple transformations and an elementary estimation of the remainder, that

$$T_{-\rho,-\sigma_1} = T^0_{-\rho,-\sigma_1} + \varepsilon\left[\frac{1}{2\sigma_1} - \left(\frac{1}{2}b^1_{\varkappa'}g_1^{\varkappa'} + g_1^1\right)\ln\sigma_1\right] + \varepsilon^2\frac{-1}{8\sigma_1^4} + O(\varepsilon), \tag{3.5}$$

where we have used the notation

$$T^0_{-\rho,\,-\sigma_1}=\int\limits_{-\rho}^{-\sigma_1}\frac{d}{d\xi^1}\,[\eta_0^1(\xi^1)]\,\delta(\xi^1,\,\xi_0^{\alpha'}(\xi^1),\,\eta_0^{\varkappa}(\xi^1))\,d\xi^1. \qquad (3.6)$$

Now consider the integral in (3.4) on the interval $-\sigma_1 \leq \xi^1 \leq 0$; the variable change (IV.5.1) yields

$$T_{-\sigma_1,\,0}=\mu^2\int\limits_{-\omega_1}^{0}\frac{d}{du^1}\,[v^1(u^1,\,\mu)]\,\frac{du^1}{\psi^1(u^1,\,u^{\alpha'}(u^1,\,\mu),\,v^{\varkappa}(u^1,\,\mu),\,\mu)}; \qquad (3.7)$$

here $\omega_1 = \sigma_1/\mu$, the functions $\psi^1(u^\alpha, v^\kappa, \mu)$ are defined in (IV.5.3) and (IV.5.2), and the functions

$$u^i=u^i(u^1,\,\mu),\quad i=2,\,\ldots,\,k,\qquad v^j=v^j(u^1,\,\mu),\quad j=1,\,\ldots,\,l, \qquad (3.8)$$

in the new coordinates describe the part (3.2). For $-\sigma_1 < \xi^1 \leq 0$, the functions (3.2) have the asymptotic representations (IV.5.18); hence we can use (IV.5.15) for the functions (3.8). It is clear that

$$\frac{1}{\psi^1(u^1,\,u^{\alpha'}(u^1,\,\mu),\,v^{\varkappa}(u^1,\,\mu),\,\mu)}=1-\mu g_1^1u^1+O(\mu^2(u^1)^2).$$

Employment of these representations in (3.7) and estimation of the remainder term now yield

$$T_{-\sigma_1,\,0}=\mu^2\int\limits_{-\omega_1}^{0}\frac{d}{du^1}\,[v_0^1(u^1)+\mu v_1^1(u^1)]\,(1-\mu g_1^1u^1)\,du^1+O(\varepsilon). \qquad (3.9)$$

In the light of the expansion (IV.4.9) for the trajectory of the degenerate system, we can write (3.9) as

$$\begin{aligned}T_{-\sigma_1,\,0}&=\mu^2\int\limits_{-\omega_1}^{0}\frac{d}{du^1}\Big[-(u^1)^2+\\&+\mu\Big(\frac{2}{3}b^1_{\varkappa'}g_1^{\varkappa'}+c_1^1-d_1^1-r^1_{\alpha'}k^{\alpha'}\Big)(u^1)^3\Big]\,(1-\mu g_1^1u^1)\,du^1+\\&+\mu^2\int\limits_{-\omega_1}^{0}\frac{d}{du^1}\,[z_0(u^1)]\,(1-\mu g_1^1u^1)\,du^1+\mu^3\int\limits_{-\omega_1}^{0}\frac{d}{du^1}\Big[v_1^1(u^1)-\\&-\Big(\frac{2}{3}b^1_{\varkappa'}g_1^{\varkappa'}+c_1^1-d_1^1-r^1_{\alpha'}k^{\alpha'}\Big)(u^1)^3\Big]\,(1-\mu g_1^1u^1)\,du^1+O(\varepsilon); \qquad (3.10)\end{aligned}$$

the first term on the right here will be denoted by $T^0_{-\sigma_1,0}$. Transforming back to the variable of integration ξ^1, we obtain

$$T^0_{-\sigma_1,0} = \int_{-\sigma_1}^{0} \frac{d}{d\xi^1}\left[-(\xi^1)^2 + \left(\frac{2}{3} b^1_{\varkappa'} g_1^{\varkappa'} + c_1^1 - d_1^1 - r^1_{\alpha'} k^{\alpha'}\right)(\xi^1)^3\right](1 - g_1^1 \xi^1)\, d\xi^1 =$$

$$= \int_{-\sigma_1}^{0} \frac{d}{d\xi^1}[\eta_0^1(\xi^1)]\,\delta(\xi^1, \xi_0^{\alpha'}(\xi^1), \eta_0^{\varkappa}(\xi^1))\, d\xi^1 + O(\varepsilon). \qquad (3.11)$$

We now integrate by parts the second term on the right of (3.10) and, using the representation (IV.5.7) for $u^1 \to -\infty$, we regularize the integral (see Chapter II, Sec. 13) to obtain

$$\mu^2 \int_{-\omega_1}^{0} \frac{d}{du^1}[z_0(u')](1 - \mu g_1^1 u^1)\, du^1 =$$

$$= \mu^2 z_0(0) - \frac{\mu^3}{2\sigma_1} + \frac{\mu^6}{8\sigma_1^4} + \frac{1}{2} g_1^1 \mu^3 \ln\frac{1}{\mu} + \frac{1}{2} g_1^1 \mu^3 \ln \sigma_1 + O(\varepsilon).$$

We integrate by parts the third term on the right of (3.10) and use the representation (IV.5.13) for $u^1 \to -\infty$:

$$\mu^3 \int_{-\omega_1}^{0} \frac{d}{du^1}\left[v_1^1(u^1) - \left(\frac{2}{3} b^1_{\varkappa'} g_1^{\varkappa'} + c_1^1 - d_1^1 - r^1_{\alpha'} k^{\alpha'}\right)(u^1)^3\right](1 - \mu g_1^1 u^1)\, du^1 =$$

$$= \frac{1}{2} b^1_{\varkappa'} g_1^{\varkappa'} \mu^3 \ln\frac{1}{\mu} + \frac{1}{2} b^1_{\varkappa'} g_1^{\varkappa'} \mu^3 \ln \sigma_1 + O(\varepsilon).$$

Finally, we obtain

$$T_{-\sigma_1,0} = T^0_{-\sigma_1,0} + z_0(0)\,\varepsilon^{2/3} - \frac{\varepsilon}{2\sigma_1} + \frac{\varepsilon^2}{8\sigma_1^4} +$$

$$+ \frac{1}{6}(g_1^1 + b^1_{\varkappa'} g_1^{\varkappa'})\,\varepsilon \ln\frac{1}{\varepsilon} + \frac{1}{2}(g_1^1 + b^1_{\varkappa'} g_1^{\varkappa'})\,\varepsilon \ln \sigma_1 + O(\varepsilon). \qquad (3.12)$$

Similar reasoning applied to the integral in (3.4) on the interval $0 \le \xi^1 \le \sigma_2$ shows that

$$T_{0,\sigma_2} = \mu^2 \int_0^{\omega_2} \frac{d}{du^1}[v_0^1(u^1) + \mu v_1^1(u^1)](1 - \mu g_1^1 u^1)\, du^1 + O(\varepsilon);$$

here $\omega_2 = \sigma_2/\mu$ [cf. (3.9)]. We rewrite this formula as

$$T_{0,\sigma_2} = \mu^2 \int_0^{\omega_2} \frac{d}{du^1}[v_0^1(u^1)]\, du^1 -$$

$$- \mu^3 g_1^1 \int_0^{\omega_2} \frac{d}{du^1}[v_0^1(u^1)]\, u^1\, du^1 + \mu^3 \int_0^{\omega_2} \frac{d}{du^1}[v_1^1(u^1)]\, du^1 + O(\varepsilon).$$

We integrate by parts, using (IV.5.7) and (IV.5.3) for $u^1 \to +\infty$, regularize the integral, and, after some elementary transformations, obtain

$$T_{0,\sigma_2} = [\Omega_0 - z_0(0)]\,\varepsilon^{2/3} - \frac{\varepsilon}{\sigma_2} -$$
$$-\frac{1}{3}(d_1^1 + r_{\alpha'}^1 \Omega_0^{\alpha'})\,\varepsilon \ln \frac{1}{\varepsilon} - (d_1^1 + r_{\alpha'}^1 \Omega_0^{\alpha'})\,\varepsilon \ln \sigma_2 + O(\varepsilon). \qquad (3.13)$$

We have the asymptotic representations (IV.6.10) for the functions (3.2) on the interval $\sigma_2 \leq \xi^1 \leq \rho$. Considering the integral in (3.4) on the interval $\sigma_2 \leq \xi^1 \leq \rho$ and using the representations (IV.6.10) in the integrand, we obtain the following result after some elementary transformations:

$$T_{\sigma_2,\rho} = \frac{\varepsilon}{\sigma_2} + (d_1^1 + r_{\alpha'}^1 \Omega_0^{\alpha'})\,\varepsilon \ln \sigma_2 + O(\varepsilon). \qquad (3.14)$$

To find $T_{-\rho,\rho}$, it remains to use (3.5), (3.12), (3.13), and (3.14) in (3.3):

$$T_{-\rho,\rho} = T^0_{-\rho,0} + \Omega_0 \varepsilon^{2/3} + \left(\frac{1}{6}Q - \frac{1}{3}D - \frac{1}{3}K\right)\varepsilon \ln \frac{1}{\varepsilon} + O(\varepsilon)\,; \qquad (3.15)$$

here $T^0_{-\rho,0} = T^0_{-\rho,-\sigma_1} + T^0_{-\sigma_1,0}$ [see (3.6) and (3.11)] is the time taken by the phase point of the degenerate system (IV.1.2) to traverse the part of $u_1(\delta S_1)$ between S_1' and $S_1 + \delta S_1$. The constant Ω_0 is independent of the specific form of the system (IV.1.1); it is a universal constant (see Chapter II, Sec. 9). Furthermore,

$$Q = g_1^1 + b_{\varkappa'}^1 g_1^{\varkappa'}, \quad D = d_1^1, \quad K = r_{\alpha'}^1 \Omega_0^{\alpha'}; \qquad (3.16)$$

invariant expressions for these quantities are given in Chapter IV [see (IV.3.12), (IV.7.5), and (IV.7.6)].

The coefficients (3.16) in (3.15) clearly depend on values of the right sides in (IV.1.1) and certain of their derivatives at $S_1 + \delta S_1$. But, with accuracy up to quantities of order ε, the time $T_{-\rho,\rho}$ remains the same if, instead of the coefficients (3.16) corresponding to the junction point $S_1 + \delta S_1$, we use quantities corresponding to the point S_1. We denote these quantities by Q_1, D_1, and K_1.

We have thus obtained an asymptotic formula for the time $T^1_{-\rho,\rho}$ taken by the phase point of the nondegenerate system (IV.1.1) to traverse the part $Z_\varepsilon(S_1 + \delta S_1)$

$$T^1_{-\rho,\rho} = T^{1,0}_{-\rho,0} + \Omega_0 \varepsilon^{2/3} + \left(\frac{1}{6}Q_1 - \frac{1}{3}D_1 - \frac{1}{3}K_1\right)\varepsilon \ln \frac{1}{\varepsilon} + O(\varepsilon); \qquad (3.17)$$

here $T^{1,0}_{-p,0}$ is the traversal time of the phase point of the degenerate system (IV.1.2) on the part $(S_1', S_1 + \delta S_1)$ of the trajectory $u_1(\delta S_1)$. The calculation is similar for the part $Z_\varepsilon(S_2 + \delta S_2)$ of Z_ε; it yields

$$T^2_{-\rho,\rho} = T^{2,0}_{-\rho,0} + \Omega_0 \varepsilon^{2/3} + \left(\frac{1}{6} Q_2 - \frac{1}{3} D_2 - \frac{1}{3} K_2\right) \varepsilon \ln \frac{1}{\varepsilon} + O(\varepsilon). \tag{3.18}$$

We now find the time τ_1 taken in the space Y^l to pass from the point Δ_1 to the plane $N_2\tilde{S}_2$ at the constant velocity $g(P_1)$. The plane $\tilde{S}_2$ is determined in Y^l by the covariant vector ${}_2w = ({}_2w_1, \ldots, {}_2w_l)$ corresponding to the junction point S_2; the components of this vector have already been calculated [see (1.3) and (1.4)]. It is easily verified that $N_2\tilde{S}_2$ (the image of the plane $\tilde{S}_2$ under the mapping N_2) is determined by the covariant vector ${}_2w' = ({}_2w_1', \ldots, {}_2w_l')$, and

$${}_2w'_\alpha = q^\beta_\alpha {}_2w_\beta,$$

where $\| q^\alpha_\beta \|$ is the matrix of the transformation N_2^{-1}. Elementary calculations show that

$$\tau_1 = \frac{({}_2w', \Delta_1)}{({}_2w', g(P_1))} + O(\varepsilon).$$

But it is easily seen that

$$({}_2w', g(P_1)) = ({}_2w, N_2^{-1} g(P_1)) = ({}_2w, g(S_2)) = 1,$$
$$({}_2w', \Delta_1) = ({}_2w, N_2^{-1} \Delta_1);$$

hence

$$\tau_1 = ({}_2w, N_2^{-1} \Delta_1) + O(\varepsilon). \tag{3.19}$$

A similar calculation yields the time τ_2 for passage from Δ_2 to the plane $N_1\tilde{S}_1$ with constant velocity $g(P_2)$

$$\tau_2 = ({}_1w, N_1^{-1} \Delta_2) + O(\varepsilon). \tag{3.20}$$

Now, using (3.1), (3.17)-(3.20), and Theorem 1, we can obtain an asymptotic formula for the relaxation-oscillation period.

Theorem 2. The period T_ε of the trajectory Z_ε of the nondegenerate system (IV.1.1) has the asymptotic representation

$$T_\varepsilon = T_0 + \Delta T^1_\varepsilon + \Delta T^2_\varepsilon + O(\varepsilon),$$

where T_0 is the period of the trajectory Z_0 of the degenerate system (IV.1.2), and

$$\Delta T_\varepsilon^1 = \Omega_0 \varepsilon^{2/3} + \left(\frac{1}{6} Q_1 - \frac{1}{3} D_1 - \frac{1}{3} K_1\right) \varepsilon \ln \frac{1}{\varepsilon} - \tau_1,$$
$$\Delta T_\varepsilon^2 = \Omega_0 \varepsilon^{2/3} + \left(\frac{1}{6} Q_2 - \frac{1}{3} D_2 - \frac{1}{3} K_2\right) \varepsilon \ln \frac{1}{\varepsilon} - \tau_2;$$

these quantities can be calculated without solving the non-degenerate system (IV.1.1).

REFERENCES

1. J. Hadamard, *Le problème de Cauchy et les équations aux dérivées partielles linéaires hyperboliques*, Hermann, Paris (1932). See also: *Lectures on Cauchy's Problems in Linear Hyperbolic Differential Equations*, Dover, New York (1952).
2. A. A. Andronov and A. A. Vitt, Discontinuous periodic solutions and the Abraham-Bloch theory of multivibrators, *Dokl. AN SSSR*, No. 8, 189-192 (1930). See also: A. A. Andronov, *Collected works*, Izd. AN SSSR, 65-69 (1956).
3. A. A. Andronov, A. A. Vitt, and S. E. Khaikin, *Theory of Oscillations* [in Russian], Second edition, improved and extended by N. A. Zheleztsov, Fizmatgiz, Moscow (1959).
4. D. V. Anosov, Averaging in systems of ordinary differential equations with rapidly oscillating solutions, *Izv. AN SSSR, Ser. Matem.*, *24*, No. 5, 721-742 (1960).
5. D. V. Anosov, Limit cycles of systems of differential equations with a small parameter multiplying the highest-order derivatives, *Matem. Sb.*, *50*, No. 3, 299-334 (1960).
6. M. A. Belyaeva, The approximate solution of a system of ordinary differential equations with a small parameter multiplying certain derivatives, *Dokl. AN SSSR*, *189*, No. 6, 1167-1170 (1969).
7. N. N. Bogolyubov and Yu. A. Mitropol'skii, *Asymptotic Methods in the Theory of Nonlinear Oscillations* [in Russian], Nauka, Moscow (1974).
8. V. F. Butuzov, A. B. Vasil'eva, and M. V. Fedoryuk, Asymptotic methods in the theory of ordinary differential equations, *Progress in Mathematics, Vol. 8: Mathematical Analysis*, Plenum Press, New York (1970), pp. 1-82.
9. W. R. Wasow, Singular-perturbation methods for nonlinear oscillations, *Proceedings of a Symposium on Nonlinear Circuit Analysis*, Polytechn. Inst. of Brooklyn (1953), pp. 75-98.

10. W. R. Wasow, *Asymptotic Expansions of Solutions of Ordinary Differential Equations*, Krieger, Huntington, New York (1966).
11. B. Van der Pol, On relaxation Oscillations, *Philos. Mag., (7), 2,* No. 11, 978-992 (1926).
12. A. B. Vasil'eva, Differential equations with small parameters multiplying derivatives, *Matem. Sb., 31,* No. 3, 587-644 (1952).
13. A. B. Vasil'eva, Asymptotic expansions of solutions of some problems for ordinary nonlinear differential equations with small parameters multiplying leading derivatives, *Uspekhi Matem. Nauk, 18,* No. 3, 15-86 (1963).
14. A. B. Vasil'eva and V. F. Butuzov, *Asymptotic Expansions of Solutions of Singularly Perturbed Equations* [in Russian], Nauka, Moscow (1973).
15. V. M. Volosov and B. I. Morgunov, *The Averaging Method in the Theory of Nonlinear Oscillatory Systems* [in Russian], Izd-vo MGU, Moscow (1971).
16. I. S. Gradshtein and I. M. Ryzhik, *Tables of Integrals, Sums, Series, and Products* [in Russian], Nauka, Moscow (1971).
17. N. G. De Bruijn, *Asymptotic Methods in Analysis*, North-Holland, New York (1970).
18. A. A. Dorodnitsyn, Asymptotic solution of Van der Pol's equation, *Prikl. Matem. i Mekhan., 11,* No. 3, 313-328 (1947).
19. N. A. Zheleztsov, The theory of discontinuous oscillations in second-order systems, *Izv. Vysshikh Uchebn. Zavedenii. Radiofizika, 1,* No. 1, 67-78 (1958).
20. N. A. Zheleztsov and L. V. Rodygin, The theory of the symmetric multivibrator, *Dokl. AN SSSR, 81,* No. 3, 391-392 (1951).
21. E. Kamke, *Manual of Ordinary Differential Equations* [Russian translation], Nauka, Moscow (1971). [German edition, Chelsea, New York, 1967.]
22. M. L. Cartwright, Van der Pol's equation for relaxation oscillations, in *Contributions to the Theory of Nonlinear Oscillations*, S. Lefschetz, ed., Princeton University Press (1952), pp. 3-18.
23. E. A. Coddington and N. Levinson, *The Theory of Ordinary Differential Equations*, McGraw-Hill, New York (1955).
24. E. T. Copson, *Asymptotic Expansions*, Cambridge University Press, New York (1968).
25. Ph. le Corbeiller, *Les Systemes autoentretenues et les Oscillations de relaxation*, Hermann, Paris (1931).

26. J. D. Cole, *Perturbation Methods in Applied Mathematics*, Blaisdell, Lexington, Massachussetts (1968).
27. N. M. Krylov and N. N. Bogolyubov, *Introduction to Nonlinear Mechanics* [in Russian], Izd-vo AN UkrSSR, Kiev (1937).
28. J. LaSalle, Relaxation oscillations, *Quart. J. Appl. Math.*, *7*, No. 1, 1-19 (1949).
29. N. Levinson, Perturbations of discontinuous solutions of nonlinear systems of differential equations, *Acta Math.*, *82*, Nos. 1 and 2, 71-106 (1951).
30. S. Lefschetz, *Differential Equations: Geometric Theory*, Wiley, New York (1963).
31. G. S. Makaeva, The asymptotic properties of solutions of differential equations with a small parameter; fast-motion systems, including Hamiltonian systems, *Izv. AN SSSR, Ser. Matem.*, *25*, No. 5, 685-716 (1961).
32. Yu. A. Mitropol'skii, *Problems in the Asymptotic Theory of Nonstationary Oscillations* [in Russian], Nauka, Moscow (1964).
33. Yu. A. Mitropol'skii, *The Averaging Method in Nonlinear Mechanics* [in Russian], Naukova Dumka, Kiev (1971).
34. E. F. Mishchenko, Asymptotic calculation of periodic solutions of systems of differential equations with small parameters multiplying derivatives, *Izv. AN SSSR, Ser. Matem.*, *21*, No. 5, 627-654 (1957).
35. E. F. Mishchenko, Asymptotic theory of relaxation oscillations in second-order systems, *Matem. Sb.*, *44*, No. 4, 457-480 (1958).
36. E. F. Mishchenko, Asymptotic methods in the theory of relaxation oscillations, *Uspekhi Matem. Nauk*, *14*, No. 6, 229-236 (1959).
37. E. F. Mishchenko, *Differential Equations Containing a Small Parameter and Relaxation Oscillations*, University of Michigan (1964).
38. E. F. Mishchenko and L. S. Pontryagin, Almost discontinuous periodic solutions of differential equation systems, *Dokl. AN SSSR*, *102*, No. 5, 889-891 (1955).
39. E. F. Mishchenko and L. S. Pontryagin, Proof of certain asymptotic formulas for solutions of differential equations containing small parameters, *Dokl. AN SSSR*, *120*, No. 5, 967-969 (1958).
40. E. F. Mishchenko and L. S. Pontryagin, Derivation of some asymptotic estimates of solutions of differential equations with small parameters multiplying derivatives, *Izv. AN SSSR, Ser. Matem.*, *23*, No. 5, 643-660 (1969).

41. E. F. Mishchenko and L. S. Pontryagin, Differential equations with a small parameter attached to the higher derivatives and some problems in the theory of oscillations, *IRE Trans. Circuit Theory, 7,* No. 4, 527-535 (1960).
42. E. F. Mishchenko and L. S. Pontryagin, Relaxation oscillations and differential equations containing a small parameter with the senior derivative. The Golden Jubilee Commemoration Volume (1958-1959), Calcutta Mathematical Society, Calcutta (1963), pp. 141-150.
43. L. S. Pontryagin, Asymptotic properties of solutions of systems of differential equations with a small parameter multiplying leading derivatives, *Izv. AN SSSR, Ser. Matem., 21,* No. 5, 605-626 (1957).
44. L. S. Pontryagin, *Systems of Ordinary Differential Equations with a Small Parameter Multiplying Higher Derivatives,* Trans. of the Third All-Union Mathematics Symposium, Vol. III [in Russian], Izd. AN SSSR, 570-577 (1958).
45. L. S. Pontryagin, *Ordinary Differential Equations* [in Russian], Nauka, Moscow (1965).
46. L. S. Pontryagin and L. V. Rodygin, Approximate solution of a system of ordinary differential equations with a small parameter multiplying derivatives, *Dokl. AN SSSR, 131,* No. 2, 255-258 (1960).
47. L. S. Pontryagin and L. V. Rodygin, A periodic solution of a system of ordinary differential equations with a small parameter multiplying derivatives, *Dokl. AN SSSR, 132,* No. 3, 537-540 (1960).
48. S. S. Pul'kin and N. Kh. Rozov, The asymptotic theory of relaxation oscillations in systems with one degree of freedom. I. Calculation of phase trajectories, *Vestn. Mosk. Un-ta, Ser. Matem., Mekhan.,* No. 2, 70-82 (1964).
49. L. V. Rodygin, Existence of an invariant torus for a system of ordinary differential equations containing a small parameter, *Izd. Vysshikh Uchebn. Zavedenii. Radiofizika, 3,* No. 1, 116-129 (1960).
50. N. Kh. Rozov, Asymptotic calculation of almost discontinuous periodic solutions of second-order differential equation systems, *Dokl. AN SSSR, 145,* No. 1, 38-40 (1962).
51. N. Kh. Rozov, Asymptotic theory of relaxation oscillations in systems with one degree of freedom. II. Calculation of a limit-cycle period, *Vestn. Mosk. Un-ta, Ser. Matem., Mekhan.,* No. 3, 56-65 (1964).

52. N. Kh. Rozov, Asymptotic calculation of almost discontinuous periodic solutions describing relaxation oscillations in systems with one degree of freedom, International Mathematical Congress, Brief Communications, Section 6 [in Russian], Moscow (1966), pp. 45-46.
53. Ya. Sibuya, On perturbations of discontinuous solutions of ordinary differential equations, *Natur. Sci. Rept. Ochanomizu Univ., 11,* No. 1, 1-18 (1960).
54. A. S. Sidorov, *The Theory and Design of Nonlinear Impulse Systems for Tunnel Diodes* [in Russian], Sov. Radio, Moscow (1971).
55. J. J. Stoker, *Nonlinear Vibrations in Mechanical and Electrical Systems,* Wiley, New York (1950).
56. A. N. Tikhonov, Dependence of solutions of differential equations on small parameters, *Matem. Sb., 22,* No. 2, 193-204 (1948).
57. A. N. Tikhonov, Systems of differential equations containing small parameters multiplying derivatives, *Matem. Sb., 31,* No. 3, 575-586 (1952).
58. M. Urabe, Numerical investigation of periodic solutions of a Van der Pol equation, Trans. of an International Symposium on Nonlinear Oscillations. II [in Russian], Izd. AN UkrSSR, Kiev (1963), pp. 267-276.
59. D. A. Flanders and J. J. Stoker, The limit case of relaxation oscillations, in *Studies in Nonlinear Vibration Theory,* New York University (1946), pp. 51-64.
60. L. Flatto and N. Levinson, Periodic solutions of singularly perturbed systems, *J. Rational Mech. Anal., 4,* No. 6, 943-950 (1955).
61. J. Haag, Etude asymptotique des oscillations de relaxation, *Ann. Sci. Ecole Norm. Supp. (3) 60,* 35-64 (1943).
62. J. Haag, Exemples concrets d'étude asymptotique d'oscillation de relaxation, *Ann. Sci. Ecole Norm. Supp. (3) 61,* 65-111 (1944).
63. J. Haag, *Les mouvements vibratoires,* Press Univ. de France, Paris, Vol. I (1952), Vol. II (1955).
64. M. A. Shishkova, A system of differential equations with a small parameter multiplying higher derivatives, *Dokl. AN SSSR, 209,* No. 3, 576-579 (1973).
65. A. Erdelyi, *Asymptotic Expansions,* Dover, New York (1961).
66. E. Jahnke, F. Emde, and F. Lösch, *Tables of Functions with Formulas and Curves,* Dover, New York (1945).

INDEX

Airy's equation, 69
Almost-discontinuous periodic solutions, 139-170
  asymptotic calculation of trajectory for, 203-209
  existence of, 139-142, 203-209
  in systems of arbitrary order, 199-216
  uniqueness of, 139-142
Approximate solution, defined, 30
Approximations
  asymptotic: *see* Asymptotic approximations
  zeroth: *see* Zeroth approximation
Arbitrary-order asymptotic solutions, 171-198
Arbitrary-order systems, 16-18
  almost-discontinuous periodic solutions in, 199-216
  asymptotic calculation of solutions for, 171-198
Asymptotic approximations, 31-32
  vs. actual trajectories at end of junction section, 99-103
Asymptotic approximations (continued)
  vs. actual trajectories in immediate vicinity of junction point, 77-82
  partial sum of series from, 51-52
  of trajectory at beginning of junction system, 179-187
  of trajectory at end of junction section, 193-196
  of trajectory in fast-motion part, 107-111
  of trajectory for initial slow-motion and drop parts, 132-137
  for trajectory in neighborhood of junction point, 187-192
  for trajectory of periodic solution, 143-144
Asymptotic expansion
  defined, 31-32
  for end of junction part of trajectory, 96-99
Asymptotic formula
  for relaxation-oscillation period, 164-168
  for trajectory approximations, 34

Asymptotic methods
  application of, 30-31
  general theory of, 29-30
Asymptotic representations
  of initial drop part, 123
  proof of, 52-56
Asymptotic representations for drop part, proof of, 125-132
Asymptotic representations for fast-motion part, derivation of, 111-114
Asymptotic representations for junction part, proof of, 103-107
Asymptotic series, for coefficients of expansion near junction point, 82-88
Asymptotic solutions
  arbitrary-order, 174-198
  second-order, 39-137
Auto-oscillations, relaxation-oscillations and, 7

Bendixon's criteria, 21-22
Bessel equation, 69
Bessel function, modified, 69
Bifurcation value, 16

Closed trajectory, isolated and stable properties of, 200
Coefficients of expansion, asymptotic series for, 82-88

Degenerate system
  arbitrary-order system and, 17
  closed phase trajectories of, 37, 139
  defined, 9
Degenerate system (continued)
  function system as solution of, 27
  nondegenerate system and, 39
  periodic solution for, 199
  phase trajectory of, 173
  solutions of, 24-29, 199
  trajectories of solutions in, 10-11, 29
Dependence
  continuous, 2
  discontinuity in, 4
  smooth, 1-3
  types of, 1-37
Discontinuity, in dependence, 4
Discontinuous periodic solution, defined, 140
Discontinuous solution
  defined, 29
  phase trajectories and, 45
  zeroth order in, 173-176
Discontinuous system, trajectory of, 29
Displacement vector, 197-198
Dividing solution, for Riccati's equation, 71
Dorodnitsyn's formula, 168-170
Drop-off, defined, 13
Drop part of trajectory
  asymptotic approximations of, 119-125
  defined, 48
  initial, 48, 123
  proof of asymptotic representations of, 125-132
  special variables for, 114-119
Drop point
  defined, 173

Drop point (continued)
junction point and, 29
Drop point following junction point, 45
Drop time, calculation of, 157-164

Equilibrium position, solutions as, 26
Expansion
asymptotic, 31-32
coefficients of, 82-88

Fast motion, defined, 16-20
Fast-motion equations
defined, 16, 44
exponentially stable period solution for, 36
Hamiltonian and, 37
scalar form of, 26
Fast-motion equation system, stationary solutions of, 35
Fast-motion part of trajectory
asymptotic approximations of trajectory on, 107-111
defined, 45, 48, 174
Fast-motion time
calculation of, 156-157
fast-motion equation and, 44
Fast variable, defined, 16, 18
Finite time interval, solution dependence in, 4
Frugauer generators, 8
Function system, as solution of degenerate system, 27

Generalized integral, 90, 95

Improper integrals, regularization of, 89-95

Infinite time interval, solution dependence on, 3-4
Initial drop part
asymptotic representations of, 123
defined, 48
Initial fast-motion part
asymptotic approximation for, 134
defined, 48
Initial junction section, asymptotic representations vs. actual trajectories in, 62-67
Initial slow-motion and drop parts, asymptotic approximations of, 132-137
Integral, generalized, 90, 95

Junction, trajectory on initial part of, 60-62
Junction part of trajectory
asymptotic expansions for end of, 96-99
defined, 48
proof of asymptotic representations of, 103-107
Junction point
asymptotic approximations vs. actual trajectories in immediate vicinity of, 77-82
asymptotic approximations for trajectory in neighborhood of, 187-193
asymptotic series for coefficients of expansion near, 82-88

Junction point (continued)
  defined, 29, 42, 173
  local coordinates in neighborhood of, 56-59, 176-179
  trajectory in neighborhood of, 72-76
Junction section
  asymptotic approximations vs. actual trajectories at end of, 99-103
  asymptotic approximations of trajectory at beginning of, 179-187
  asymptotic approximation of trajectory at end of, 193-196
  special variables for, 67-68
Junction time, calculation of, 145-156

Limit cycle, for Van der Pol's equation, 35
Lyapunov function, 205

Mapping, of trajectories, 11-12
Multidimensional system, periodic solutions for, 36
Multivibrators, symmetric, 8
Mutual inductance, in Van der Pol's equation, 6-8

Newton-Leibnitz formula, 90
Nondegenerate points, defined, 40
Nondegenerate system
  closed trajectory of, 141
  defined, 39
  phase velocity of, 11

Nonregular point
  defined, 40, 172
  nondegeneracy of, 41
Numerical methods, approximate solution in, 30

Ordinary points, defined, 40

Parameter, asymptotic expansions of solutions with respect to, 29-34
Periodic solutions
  Dorodnitsyn's formula and, 168-170
  drop time and, 157-164
  exponentially stable, 36
  fast-motion time and, 156-157
  junction time and, 145-156
  slow-motion time and, 144-145
  Van der Pol's equation and, 168-170
Periodic solution trajectory, asymptotic approximations for, 143-144
Phase plane, trajectory and, 47
Phase point, stable equilibrium and, 16, 24
Phase portrait
  for fast and slow systems, 21
  of second-order system, 13
  of Van den Pol's equation, 13
Phase trajectories
  assumptions and definitions in, 39-40
  discontinuous solution and, 45
  zeroth approximation of, 45

Phase-velocity vector
infinite first component of, 15
second component of, 18
Poincaré's theorem, 1-3

Rapid-motion equation system, defined, 18
Rayleigh's equation, 6-7
Regular parts, defined, 40
Regular point, defined, 40
Relaxation motion, defined, 20
Relaxation-oscillation period, asymptotic formula for, 164-168
Relaxation oscillations
asymptotic formula for period of, 209-216
asymptotic theory and, 35
auto-oscillations and, 7
defined, 9, 13, 16
Riccati's equation, 68-72
dividing solution for, 71

Second-order systems
almost-discontinuous periodic solutions of, 139-170
asymptotic calculations and, 34, 39-137
fast and slow motions in, 9-16
phase portrait of, 13
Slow motion, in arbitrary-order systems, 16-20
Slow motion part
asymptotic approximations on, 48-52
defined, 50, 174
in phase-velocity systems, 48
proof of asymptotic representations of, 52-56
Slow-motion time, calculation of, 144-145
Slow variable, defined, 16-20

Small parameters
dependence of solutions on, 1-9
equations with, 4-9
Smooth dependence, 1-3; *see also* Dependence
Solutions
asymptotic calculation of, 171-198
asymptotic expansions of with respect to parameter, 29-34
differences between, 10
differentiability of with respect to parameter, 2-3
Stable equilibrium
defined, 13-14
phase point and, 24
Stable part, defined, 13, 16, 42
Stable region, defined, 26
Stationary solutions, of fast-motion equation system, 35
Surge, defined, 13

Trajectory
asymptotic approximations on initial part of junction, 60-62
asymptotic approximations on slow-motion parts of, 48-52
asymptotic approximations for trajectory in neighborhood of junction point, 72-76
asymptotic representation of, vs. actual, 62-67
of discontinuous system, 29
drop part of: *see* Drop part of trajectory
on initial part of junction, 60-62
mapping of, 11

Trajectory approximations, asymptotic formulas for, 34
Trajectory at beginning of junction system, asymptotic approximations for, 179-187
Trajectory at end of junction section, asymptotic approximations for, 193-196
Trajectory in neighborhood of junction point, asymptotic approximations for, 187-193

Unstable part, defined, 13, 42

Vacuum tube oscillator, equation for, 7
Van der Pol's equation, 5-7, 10, 14
- limit cycle for, 35
- periodic solutions and, 168-170
- phase portrait of, 13

Zeroth approximation
- defined, 31-32
- discontinuous solution and, 173-176
- Riccati's equation and, 68
- of trajectory in neighborhood of junction point, 75